PHYTOCHEMICALS
A PRACTICAL APPROACH

Dr. Rameshwar Dayal

Formerly

Scientist-F and Head (Retired), Chemistry Division
Forest Research Institute (ICFRE)
Dehra Dun, India

TECHSAR™

NEW DELHI

Published and Printed by

Techsar Pvt. Ltd.
4435-36/7, Ansari Road, Daryaganj
New Delhi–110 002 (India)
E-mail: tech_info@techsarworld.com
Website: www.techsarworld.com

ISBN: 978-93-85909-82-5

Reprint 2025

Preface

Humans, animals and plants exist on the earth since time immemorial. The survival of the first two on the earth is not possible without plants. The reason being that plants provide oxygen, food, shelter, wood, fuel wood, medicine and many useful products for their survival on the earth. India comprises around 45,500 plants species which occur in the forests and on the other areas on the earth. Forests provide a wide range of goods and services. Goods include wood, fuel wood, fodder, and other non-wood forest products (NWFPs) (i.e., products of vegetable, mineral, and animal origin). The important services include conservation of ecosystems, maintenance of quality and supply of water at a sustained rate, release of nutrients from soils, removal of harmful chemicals from the air, climate control and prevention or reduction of the severity of avalanches, draughts, erosion, and floods. Forests also provide a wide range of economic and social benefits, such as forest products, employment, and protection of places of cultural value. Forest products are divided into two groups, namely, wood and non-wood forest products (NWFPs). Wood includes timber and fuel wood, and all other products obtained from forests are covered under NWFPs, which are defined as the goods of biological origin other than wood, derived from forests, other wooded land and trees outside forests.

Plants are the greatest organic chemists of the earth. In the nature's laboratory, they carry out complex organic syntheses with ease to manufacture complicated chemical compounds with an amazing degree of accuracy, structures and characteristics. The different properties of the plant products are due to presence of chemical compounds in them. The chemicals in the plants may be categorized as primary chemicals or secondary constituents depending on their essential role in plant metabolism and universal presence in all the plants. The common sugars, the protein amino acids, the purines and pyrimidines of nucleic acids, the chlorophyll and others are included in the category of primary constituents. The remaining chemicals of the plants belonging to different classes of compounds, viz., terpenoids, alkaloids, phenolics and miscellaneous compounds are known as secondary metabolites. Secondary compounds may be defined as those natural products, usually of plant origin, which do not function directly in the primary biochemical activities that support the growth, development and reproduction of the organism in which they occur. A number of compounds belonging to different classes of compounds have been isolated using various separation techniques (CC, TLC, PC, HPLC and others) and characterised on the basis of their spectral data (IR, UV, NMR, and MS), preparation of derivatives and degradative studies. The wood of plants consists mainly of cellulose, hemicellulose, and lignins.

The aim of the book is to provide a simple, concise and collective description about the isolation and characterization of chemical compounds from the different parts of the plants. The isolation and identification of chemical constituents from 29 plants species, analysis of essential oil from the different parts of 14 plant species and analysis of fatty oils from the seeds of 7 plant species have been described. The book consists of seven chapters giving information about chemical constituents and their structure elucidation.

The first chapter provides a very brief description of plants and forests, and the goods and services provided by them. A brief description about the main classes of secondary metabolites isolated from the different parts of the plants have been given. The references of dictionaries, encyclopedia, and books, containing the chemical constituents isolated from different parts of plants, have also been given.

The second chapter describes the basic information about the chromatography which is being used for the isolation of pure chemical compounds from the mixture of compounds present in the plants extractives. It includes a brief history, classification of chromatographic methods and separation of compounds using different chromatographic techniques, viz., adsorption chromatography, thin layer chromatography (TLC), paper chromatography (PC), gas liquid chromatography (GLC), and high performance liquid chromatography (HPLC).

The basic information about the infrared spectroscopy (IR), ultraviolet spectroscopy (UV), nuclear magnetic resonance spectroscopy (NMR), and mass spectroscopy (MS) used in the structure elucidation of pure compounds, isolated from the plant extractives, have been described in the third chapter.

Chapter 4 deals with the isolation and structure elucidation of chemical compounds from the different parts of the plant species. The details have been described under different headings, viz., plant selection, collection of the plant material, processing of plant material, extraction of the plant material, examination of the extracts, general procedure for column chromatography of extracts, structure elucidation of a pure compound, one-dimensional NMR techniques (1D), two-dimensional NMR techniques (2D), structure elucidation of triterpenoids and flavonoids, The list of plant species examined for their chemical constituents, essential oils and fatty oils have also been given.

Chapter 5 describes the isolation and structure elucidation of chemical constituents from 29 plants species. The research work has been summarized under seven headings, viz., name of plant species, family, popular name, about the plant, medicinal and other uses, research contribution and conclusion.

Chapter 6 deals with introduction about essential oils, methods of isolation of essential oils, function and general uses of essential oils and the isolation and characterization of essential oil from the different parts of 14 plant species. The research findings have been summarized under 8 headings, viz., name of plant species, method of isolation, research contribution, method of characterization, number of compounds identified, chief constituent, major compounds and conclusion.

Chapter 7 includes a brief description about fatty oils, physicochemical properties and isolation and characterization of fatty oils from the seeds of 7 plant species. The research work has been summarized under seven headings, viz., name of plant species, family, popular name, about the plant, medicinal and other uses, research contribution and conclusion.

The book will be helpful to students, researchers, academicians, scientists, industrialists, and the people who are interested in the field of natural products chemistry.

Dr. Rameshwar Dayal

about the biosynthetic routes of all secondary metabolites have been described in a number of books on Natural Products chemistry. The secondary constituents of the plants based on their biosynthetic formation in plants may be broadly classified under three categories, namely, terpenoids, phenolics, and alkaloids including other nitrogen related compounds. A number of other miscellaneous constituents form a separate category (Walton and Brown, 1999).

The secondary compounds occur in different parts of the plants, namely, bark, flowers, fruits, leaves, roots and wood. The cell wall of woody plants is made of cellulose, lignins and hemicelluloses in comparatively constant proportions. The secondary components are formed largely as extractives or extraneous material in lumens of the cells and in certain conditions (as in heartwood) small amounts arise in cell wall. Special cells or arrangements of cells present in especial genera hold large amount of extractives. These may vary in amount and composition. The secondary components, extracted from wood using solvent of increasing polarity are usually organic in nature. The exudates like gum, resin, kino, manna and dammars are present externally to cell structures. The composition of exudates and extractives are different of the same tree (Hillis, 1986).

Ambers are fossilized resins containing non-volatile terpenoids constituents which have undergone oxidation and polymerisation to form a substance resistant to chemical and microbial attack. These (Ambers) have been found to occur in the isolated regions of the world and most commonly in Strata 70 million to 2 million years old. It was traded throughout the world and known for its mystical healing and protective powers since Neolithic times (7000–8000 B.C.). It is also used in jewellery (Langenheim, 1969).

The pine trees exuded resin is of multipurpose use and specially as "naval" stores since 17th century till to date. The resin nowadays is being isolated by resin tapping methods, for example, Rill method for *Pinus roxburghii* and can also be isolated using solvent extraction of pine trees stumps. The "naval stores" are now produced in large amount and have numerous applications. The resin isolated from different sources was used by ancient people to aid the attachment of handle to weapon, tools, etc. The use oleoresin or balsam to coat sarcophagi dates back to 1400–1300 B.C. (Hillis, 1986).

Dammar resin obtained from the several genera of Dipterocarpaceae in the Malaysian region was utilised by the Pharoahs and locally for candles, making water tight boats and other household uses. The several gums obtained from *Acacia* species and other plant species consist of complex carbohydrates which in water form viscid solutions or jellies. The people from Egypt used gum arabic probably from *Acacia senegal* in the preparation of inks or paints during the ancient period (Howes, 1949). The gums are being largely used as adhesives and in confectionary. The injuries inflicted on the cambial region of *Eucalyptus* trees and some species of *Butea* and *Pterocarpus* leads to the formation of polyphenols (mainly proanthocyanidins). The kinos are astringents in nature and some may find use for the control of diarrhoea. The tannins obtained from vegetable kingdom have been used for the preparation of leather since time immemorial. *Acacia mearnsii* (bark), *Acacia catechu* (heartwood), *Terminalia chebula* (fruits) and *Rhizophora mucronata* (bark) are some of the important sources of tannins (Dayal, 2015). The sandal wood obtained from *Santalum album* and its oil was exported from India to Mediterranean countries about 1400 years ago (Burkhill, 1935). The wood from other *Santalum* species was exported from Hawaii (1810–1825) and Western Australia (1846).

Frankincense and myrrh obtained from incisions in the bark of *Boswellia* sp. and *Commiphora* sp. growing in eastern Mediterranean countries were traded by Phoenicians and others about twenty centuries ago (Howes, 1949). *Cinnamomum zeylanicum* bark from southern India and Sri Lanka was exported for flavouring and perfumery in about 300 B.C. The use of European *Cinchona* species bark for quinine for treating malaria was recorded in 1636 in Peru. The heartwood of different plant species, namely, *Caesalpinia sappan, Haematoxylon campechianum* (Logwood), *Chlorophora tinctoria, Rhus cotinus* and *Artocarpus* sp. were used for the isolation of dye during different periods starting from 13th century in the different parts of the world (Hillis, 1986; Dayal, 2015).

The secondary metabolites or secondary compounds present in the wood, different parts of the plants and exudates are accountable for their varying properties and uses. The extractives in the heartwood is much larger than in sapwood. The heartwood may be described as the inner layer of wood which, in growing tree, have ceased to contain living cells and in which the reserve materials (e.g., starch) have been removed or converted into heartwood extractives. The comprehensive biochemical and chemical information supports the view that extractives or exudates are formed *in situ* that is at the periphery of the heartwood in the bark, cambial region or periphery of lesions. The capillaries of the cell wall can also hold extractives in addition to the extractives present in cell lumens or in cavities (Hillis, 1986).

The plants existing on the earth are included into several families and genera depending on their characteristic features and a number of other factors. The plants were used by humans for food, medicine, dye, edible oil, gum and a number of other purposes. In the beginning, the utilisation of plants for food by peoples might have been started on the basis of their use by animals and other practical experiences due to lack of knowledge about the different properties of the chemical constituents of plants. The plants are the greatest organic chemists on the earth and they synthesise different classes of simple to complex compounds in laboratory of nature with meticulous details. The isolation of compounds from the plants was started about 350 years ago with the progress in scientific knowledge. Investigation on anthocyanin pigments by Robert Boyle in 1664 and isolation of some alkaloids, namely, brucine, quinine and strychnine (1817–1820) in France are some of the examples (Walton and Brown, 1999).

The structures elucidation of a large number of these compounds became feasible after the development of sophisticated instrumentation techniques and chromatography in 20th century.

The discovery of latest chromatographic techniques like adsorption chromatography, thin layer chromatography, partition chromatography, gas liquid chromatography, high performance liquid chromatography and others led to the isolation of a large number of compounds occurring in the different parts of the plants even in small amounts. The discovery of the modern techniques like infrared spectroscopy (IR), ultraviolet spectroscopy (UV), nuclear magnetic spectroscopy (NMR), mass spectrometry (MS), gas chromatography coupled mass spectrometry (GC-MS), high performance liquid chromatography (HPLC) coupled with mass spectrometry (HPLC-MS) and other modern techniques led to the structure elucidation of a large number of compounds present in the plant parts. The advancement in the organic chemistry and biochemistry was made by the contributions of scientists, such as Birch, Erdtman, Freudenberg, Geissman, Haworth, Lawrence, Mirvo, Nozoe, Robinson, Seshadri and other scientists the world over (Hillis, 1986). As a result of this more than 1,00,000 of chemical compounds were identified till the end of 20th century (Walton and Brown, 1999) and the number has increased exponentially since then. These chemicals belong to different classes of compounds like terpenoids, flavonoids, polyphenolic, tannins, carotenoids and others, have been compiled in a number of dictionaries and books, and some of these are mentioned below (Tables 1.1 and 1.2).

TABLE 1.1 Some dictionaries of chemicals from plants

S. No.	Name of Dictionary	Author	Publisher	Year of Publication
1.	Dictionary of Natural Products, 7 Volumes	J. Buckingham	Chapman and Hall, London	1994
2.	Phytochemical Dictionary	J.B. Harborne and H. Baxter	Taylor and Francis, London	1993
3.	Dictionary of Terpenoids	R.A. Hall and J.D. Connolly	Chapman and Hall, London	1991
4.	Dictionary of Alkaloids, 2 Volumes	I. W. Southon and J. Buckingham	Chapman and Hall, London	1989
5.	Encyclopaedia of Terpenoids, 2 Volumes	J.S. Glasby	Wiley, Chichester	1982

TABLE 1.2 Some books on chemicals from plants

S. No.	Title	Author	Publisher	Year of Publication
1.	Comparative Biochemistry of the Flavonoids	J.B. Harborne	Academic Press, London	1967
2.	Chemical and Botanical Guide to Lichen Products	C.F. Culberson	Univ. of North Caroline Press, Chapel Hill	1969
3.	Naurally Occurring Acetylenes	F. Bholman, T. Burkhartd & C. Zdero	Academic Press, London	1973
4.	Plant Lipid Biochemistry	C. Hitchcock & B. W. Nichols	Academic Press, London	1980
5.	Introduction to Alkaloids: A Biogenetic Approach	G. Cordell	Wiley –Interscience, New York	1981
6.	Plant Resistance to Insects	P.A. Hedin	American Chemical Society, Washington	1983
7.	Isopentenoids in Plants: Biochemistry & Function	W.D. Nes, G. Fuller, & L.S. Tsai	Marcel Dekker, New York	1984
8.	The Flavonoids: Advances in Research	J.B. Harborne	Chapman & Hall, London	1986
9.	Naturally Occurring Quinones, III. Recent Advances	R.H. Thomson	Chapman & Hall, London	1986
10.	Plant Pigments	T.W. Godwin	Academic Press, London	1988
11.	Plant Polyphenols: Vegetable Tannins Revisited	E. Haslam	University Press, Cambridge	1989
12.	Ecological Chemistry & Biochemistry of Plant Terpenoids	J.B. Harborne & F.A.T. Barberan (Editors)	Clarendon Press, Oxford	1991
13.	Plant Defence against Mammalian Herbivory	R.T. Palo & C.T. Robbins, (Editors)	CRC Press, Boca Raton	1991

14.	Introduction to Ecological Biochemistry, 4th edition	J.B. Harborne	Academic Press, London	1993
15.	The Scent of Orchids: Olfactory and Chemical Investigations	R. Kaiser	Elsevier, Amsterdam	1993

MAIN CLASSES OF PLANT SECONDARY METABOLITES

(1) Terpenoids

The terpenoids constitute a major class of compounds. Isoprene (C_5) is the basic building block of the terpenoids. The combination of head to tail union of two isoprene units led to the formation of monoterpenoids (C_{10}), three to sesquiterpenoids (C_{15}), four to diterpenoids (C_{20}), six to triterpenoids (C_{30}), and eight to tetraterpenoids (C_{40}). These occur in different parts of the plants. Leaves of *Eucalyptus* species is one of the important source of terpenoids. The essential oil obtained from *Eucalyptus* hybrid leaves was found to contain mainly monoterpenoids (Dayal et al., 1997). The leaves left out after the isolation of essential oil or air dried leaves as such contained three triterpenes, namely, ursolic acid, ursolic acid lactone, and 2-hydroxy-ursolic acid (Dayal, 1987). Ursolic acid was found to be the chief constituent (~1.20%).

The GC-MS analysis of the essential oil isolated from the heartwood of *Shorea robusta* indicated the presence of mainly sesquiterpenes (Kaur et al., 2001), while the essential oil of resin and bast also constituted mainly sesquiterpenes and a few monoterpenes (Kaur et al., 2001; Kaur et al., 2003). The chemical examination of *Dalbergia sericea* bark indicated the presence of four triterpenes namely tirucallol acetate, β-sitosterol, glutinol, and taraxerol (Aiyar et al., 1974). The essential oil isolated from the oleoresin of *Pinus insularis* and *Pinus merkusii* were found to contain mainly α-pinene (59.36%), β-pinene (31.20%), and α-pinene (57.05%), delta 3-carene (39.35%), respectively (Dayal, 1986). Bryonolic acid, a triterpenic acid was isolated and characterised in the leaves (2.80%) and bark (3.0%) for the first time in *Tetrameles nudiflora* (Dayal and Dobhal, 2001). About 20,000 terpenoids belonging to different classes have been characterised by the year 1999 and the same has increased exponentially during the last one and half decade. The identified compounds have been compiled in a number of dictionaries as given in Table 1.1.

(2) Phenolics

Numerous phenolic compounds occur in plants and several of them have been isolated in pure form and identified by using spectral data, derivatisation and degradative studies. The phenolic compounds are consisted of at least a benzene molecule containing a hydroxyl group. Aromatic compounds holding hydroxyl group directly to nucleus is called phenol and may be categorised as monohydric, dihydric, or trihydric depending on the presence of one, two, or three hydroxyl groups. The phenol containing one hydroxyl is also known as carbolic acid and may be commercially produced by cumene-phenol process. Phenol is an industrial chemical and used as an antiseptic, and disinfectant, and in the preparation of Bakelite, drugs, dyes and other products. Catechol (o-dihydroxybenzene) is a dihydric phenol and occurs in certain plant species. Phloroglucinol (1,3,5-trihydroxybenzene) is a trihydric phenol and is obtained from many plant's resins when fused with alkalis (Finar, 1973; Reprinted, 1996). Hopeaphenol is a complex phenolic compound, and was isolated from the leaves and roots of *Vateria indica* (Dayal, 1987).

(i) Aromatic Acids

The compound holding one or more carboxyl groups directly attached to the benzene nucleus are known as aromatic acids. These are known as monocarboxylic, dicarboxylic, etc., acids depending upon the presence of one, two, etc., carboxylic groups. The acids containing carboxyl group in the side-chain may be considered as aryl–substituted aliphatic acids, but are also known as aromatic acids as they exhibit aromatic properties due to presence of benzene ring. Salicylic acid (o-hydroxybenzoic acid), caffeic acid, ferulic acid and syringic acid are some of the examples, which occur in wood extractives and other parts of the plants. Salicylic acid is used in the preparation of azo-dyes, in medicine and as an antiseptic. Methyl salicylate is the major constituent of oil of wintergreen and finds use in perfumery and as a flavouring agent. Salol (phenyl salicylate) is used as an internal antiseptic. Aspirin (acetyl salicylic acid) finds use as medicine (Finar, 1973; Reprinted, 1996).

(ii) Flavonoids

The origin of flavonoids can be traced to the Latin word "flavus" meaning yellow. Flavonoid can be defined as the water-soluble polyphenolic compound consisting of 15 carbon atoms (C_6-C_3-C_6) and having 2-phenyl benzopyrone structure. The flavonoids are widely distributed in higher plants and account for the beautiful colours of flowers, fruits, vegetables, and herbs. The flavonoid-type dyes occur in different plant parts, viz., bark, fruits, flower petals, leaves, roots, and wood of higher plants, and do not appear to occur in fungi, lichens, and mosses (Dayal, 2015). Some examples of the different flavonoid compounds have been given in Table 1.3. The references of flavonoids books have been given in Table 1.2.

Presently, the flavonoids can be mainly put under the following classes:

- Flavone (4H-1-Benzopyran-4-one, 2-phenyl)
  - (i) Luteolin [4H-1-Benzopyran-4-one, 2-(3,4-dihydroxyphenyl)-5,7-dihydroxy]
- Flavonol (4H-1-Benzopyran-4-one, 3-hydroxy-2-phenyl)
  - (i) Quercetin [4H-1-Benzopyran-4-one, 2-(3,4-dihydroxyphenyl)-3,5,7-trihydroxy]
  - (ii) Kaempferol [4H-1-Benzopyran-4-one-2-(4-hydroxyphenyl)-3, 5,7 trihydroxy]
- Flavanone (4H-1-Benzopyran-4-one, 2,3-dihydro-2-phenyl)
- Isoflavone (4H-1-Benzopyran-4-one, 3-phenyl)
- Chalcone (2-Propene-1-one, 1-phenyl-3-phenyl)

(i) 2,4,4-Trihydroxychalcone [2 Propen-1-one, 1-(2,4-dihydroxylphenyl)-3 (4-hydroxyphenyl)]

- Flavan (2H-1-Benzopyran, 3, 4-didhydro-2-phenyl)
- Flavanols (2H-1-Benzopyran-3, 4 diol, 3, 4-dihydro-2-phenyl) and [2H-1-Benzopyran-3, 5, 7-triol, 2-(3,4-dihydroxyphenyl)-3, 4-dihydro]
- 3-Flavanone (2H-1-Benzopyran-3(4H)-one, 2-phenyl)
- Anthocyanins

TABLE 1.3 Some examples of the flavonoids

S. No.	Name	Examples
1.	Flavone	Apigenin, luteolin, artocarpin, eucalyptin
2.	Flavonol	Kaempferol, quercetin, isorhamnetin, myricetin
3.	Dihydroflavonols or flavanonols	Aromadendrin, fustin, pinobanksin, taxifolin
4.	Flavanone	Artoflavanone, pinocembrin, naringenin
5.	Chalcone	Stipulin, butein, carthamin
6.	Leucoanthocyanidins	Leucocyanidin, teracacidin
7.	Flavan-3-ols or catechins	(+)-Catechin, (–)-catechin, (–)-epicatechin, (+)-epicatechin
8.	Anthocyanins	Cyanin, delphinin, hirustin, peonin
9.	Isoflavones	Afromosin, genistein, prunetin, santal

(iii) Tannins

Tannins are generally organic chemical compounds consisting of C, H, and O. Some tannins contain smaller amount of other elements, such as N or P. The extensively distributed plant polyphenols containing adequate numbers of hydroxyl and other suitable functional groups (i.e., carboxyl) of fairly high molecular weight are known as tannins, and these usually occur as glycosides. Tannins have the ability to convert animal hides into leather by a process known as tanning. Vegetable tannins are of two types, namely, hydrolysable tannins and condensed tannins. The hydrolysable tannins have a polyester structure and can be easily hydrolyzed by acids or enzymes into a sugar or related polyhydric alcohol and a phenol carboxylic acid. They are further subdivided into gallotannins and ellagitannins, depending on the involvement of gallic acid or ellagic acid in esterification and their hydrolysis gives gallic acid and ellagic acid, respectively. The ellagic acid is normally isolated as its stable dilactone. Therefore, the term "hydrolysable tannin"includes both the gallotannins and the ellagitannins (Kashiwada et al., 1992). The sugar portion of the hydrolysable tannins has most often been found to be D-glucose.

Non-hydrolysable oligomeric and polymeric proanthocyanidins were classified as condensed tannins (Würdig and Woller, 1989). The condensed tannins (proanthocyanidins; molecular weight up to 20,000), in contrast, do not readily break down with acids; instead, they undergo progressive polymerization under the action of acids to yield amorphous phlobaphanes or tannin reds. These tannins are formed by condensation or polymerization of flavan-3, 4-diols and flavan-3-ol precursors. Therefore, the tannins may be classified into four prominent groups, namely, gallotannins, ellagitannins, complex tannins, and condensed tannins depending on their chemical characteristics (Anon, 2009; Khanbadaee and Ree, 2001) These are complex polyphenols and are not based on a class of compounds with a common basic structure. The molecular weight of tannins vary in the range from 500 to 3000 for hydrolysable tannins (Gallic and ellagic acid esters) and up to 20,000 for condensed tannins (proanthocyanidins) [Bate-Smith and Swain, 1962].

Tannin occurs in large number of plants and are common in Gymnosperms and Angiosperms. Tannins are situated normally in the vacuoles or surface wax of the plants. Naturally occurring tannins obtained from different parts of the plants, viz., wood, bark, fruits, and leaves, are an important non-wood forest product and are used extensively in the leather industry for converting raw hides and skins to leather and also as a mordant in dyeing. Oak galls and wattle bark are an excellent source of tannins (Dayal, 2015). Some functions of tannins in plants, uses, sources and other details have also been described (Tewari, 1994). The tannins books references containing complete information about tannins have been given in Table 1.2.

Some Functions of Tannins in Plants

(i) Biological role not fully understood.

(ii) Some contend it to be a by-product of metabolic process and has a definite function in plant.

(iii) Astringent property, prevents damage from browsing animals or creatures.
(iv) Protects trees from insect and fungal attack.
(v) Tannins or their precursors may serve as oxygen carriers in the oxidation system of plants.
(vi) Get transformed to produce coloured pigments.
(vii) Tannin-containing woods in general are dark and durable (Tewari, 1994).

Some Uses of Tannins

(i) As natural mordants in dyeing of textiles with natural dyes.
(ii) In leather industry for conversation of animals hides into leather.
(iii) To reduce the viscosity of the drill mud in deep oil-well drilling in combination with caustic soda.
(iv) In the manufacture of tannin-formaldehyde and tannin-phenol-formaldehyde resins (wattle tannins).
(v) In boiler-water treatment for the prevention and removal of boiler scale.
(vi) In the preservative treatment of fishing net.
(vii) In the flotation of certain ores.

(3) Quinones

The cyclic unsaturated diketones containing double bonds and keto groups in conjugation are known as quinones. Therefore, these are not aromatic compounds, but can be synthesised and converted to aromatic compounds. For example, o- and p-benzoquinones may be prepared from benzene. Quinones also occur in nature mainly in higher plants, fungi, and bacteria. The quinones may be put in three groups, namely, benzoquinones, naphthoquinones, and anthraquinones. The anthraquinones form the largest group and found mainly in higher plants, fungi and lichens. In, animals, some occur in insects (Coccidae only) and in feather stars (Crinoidea). Anthraquinones (often as glycosides) have been mainly isolated from heartwood, bark and roots of higher plants and sometimes in the stems, seeds and fruits. These occur in limited number of plant families. Rubiaceae, Rhamnaceae, Polygonaceae, and Verbenaceae are some of the examples. Tectoquinone, alizarin, purpurin, xanthopurpurin, emodin, morindone, aloe-emodin, and rhein are some of the examples of anthraquinone compounds isolated from plants. Emodin is probably the most widely distributed anthraquinone and a number of compounds are biosynthetic variation of this basic structure. The anthraquinones isolated from coccid insects are kermesic acid from *Kermococcus ilicis*, carminic acid from *Dactylopius coccus*, laccaic acids A, B, C, E and laccaic acid D from lac insect, *Laccifer lacca* and these have been widely used as natural dyes during the ancient period. Lac dye is still in use to a limited extent in India. Tectoquinone, one of the several quinones, is present in the teak wood (*Tectona grandis*) and contribute to its durability due to its repellent properties to termites. Alizarin is the major colouring matter of madder (*Rubia tinctorum* roots), one of the most widely used ancient natural dyeing material. The naphthoquinones (~120) are distributed sporadically and about half of them occur in higher plants, disseminated through some twenty families. These have been found to occur in bark, flowers, fruits, leaves, roots and wood. Some of the examples are lawsone from *Lawsonia inermis* leaves (Lythraceae), juglone from *Juglans regia* fruits (Juglandaceae), alkannin from *Alkanna tictoria* roots (Boraginaceae) and other compounds. The powder of the leaves of *Lawsonia inermis* (Mehndi) is being used as a dye. Lawsone was characterised as the main compound of its dye. About ninety benzoquinones have been isolated from natural sources and of these about one third each occurs in flowering plants and fungi in approximately twenty families (Thomson, 1971; Dayal, 2015).

The quinones are synthesised in nature from a few key intermediates, mainly acetate, shikimate and mevalonate by a number of reactions leading to formation of benzenoid compounds. The oxidation of a phenol in the last stage leads to the formation of quinone. The colour of naturally occurring quinones may vary from pale yellow to almost black, but have little input to natural colouring in comparison to carotenoids or anthocyanins (Thomson, 1971). The books references have been given in Table 1.2.

(4) Alkaloids

It is difficult to give a precise definition to alkaloids. The plant alkaloids still may probably be defined as the physiologically active compounds having basic properties, and containing at least one nitrogen atom in a heterocyclic ring, normally complex structures and isolated from the plants. The alkaloids are poisonous in nature and are used in small quantities in medicine. The plants alkaloids normally occur in their bark, leaf, seed or root, and usually as salts of different plant acids like acetic acid, citric acid, malic acid, oxalic acid, tartaric acid, etc. These can be isolated from different parts of the plants by extraction of the dried and powdered material with methanol. Removal of the solvent on a water bath under reduced pressure will yield the residue which on treatment with inorganic acid will separate into water soluble alkaloids solution. The mixture of alkaloids or free bases are liberated from this solution by the addition of sodium carbonate solution and may be extracted with ether, chloroform, ethyl acetate, or n-butanol depending upon the complexity of the structures of alkaloids. The pure compounds from the mixture may be isolated by chromatography (TLC, CC, PC, and HPLC) and other methods available in the literature. The structures of compounds were elucidated using spectroscopic methods (IR, UV, NMR and MS), and degradative studies.

The alkaloids are usually obtained from plants, but may also be synthesised in the laboratory (Ephedrine and Papaverine) depending on their demand. A number of alkaloids have been synthesised usually involving multi-steps synthetic route. The alkaloids isolated from plants are mostly colourless, crystalline and non-volatile solids, which are insoluble in water but soluble in organic solvents like ether, chloroform, ethanol, etc. Some exceptions are yellow coloured berberine, liquid nicotine and coniine. The later two are soluble in water. The majority of alkaloids possess bitter taste, and are laevorotatory in nature. They are commonly tertiary nitrogen compounds containing one or two nitrogen atoms, generally exist in tertiary state in a ring system and also contain oxygen. The Dragendorff's reagent and other reagents are used to detect the presence of alkaloids in the mixture and also during the paper or thin layer chromatography for their separation. The alkaloids may probably be chemically classified according to the presence of the nature of the nucleus present in molecule. The different classifications of alkaloids is also possible (Finar, 1975). More than about 10,000 structures of alkaloids have been elucidated and are given in dictionary of alkaloids. The occurrence of alkaloids is limited in higher plants and exists in about 20 per cent of angiosperm species, rarely in gymnosperms and ferns, and not in mosses or liverworts. The concentration of alkaloids is higher in fruits, roots, seeds in comparison to green leaves. The quantity of nitrogen containing compounds may be quite low in comparison to phenolic compounds in the plants due to limiting availability of nitrogen to the plants and involvement of nitrogen containing compounds in biosynthesis of alkaloids in nature (Walton and Brown, 1999). Numerous other miscellaneous secondary metabolites also occur in the plants. The books references have been given in Table 1.2.

IMPORTANCE OF SECONDARY COMPOUNDS

The secondary compounds produced in laboratory of nature by plants are essential for their survival and have a number of other uses. The secondary metabolites usually occur in the different parts of the plants as a complex mixture. The occurrence of essential oil in the leaves of *Eucalyptus* hybrid (Dayal et al., 1997) and *Cephalotaxus harringtonia* (Mendiratta et al., 2005), and azadirachtin, fatty oil and a number of other compounds in the *neem* seeds (Akhila and Rani, 1999) are some of the examples. Isolation of bryonolic acid (a triterpene) from *Tetrameles nudiflora* (leaves and bark) is an exception where only one compound is present (Dayal and Dobhal, 2001).

The secondary metabolites present in the plants depending on their chemical structures may have growth regulatory properties, involves in pollination and seed dispersal, antifungal agents, anti-herbivore agents, animal harmonal activity, symbiotic associations and other uses. The relevant details have been described in the book entitled "Chemicals from Plants" edited by N.J. Walton and D.E. Brown, in other books and in a number of research publications in different scientific journals (Harborne, 1993; Grayer and Harborne, 1994; Bailey and Mansfield, 1982; Smith and Moss 1985; Camps, 1991; Culberson, 1969; Shimizu, 1986).

Efforts have been made to give a brief introduction about the the type of chemical constituents present in the different plants. But still the huge chemical wealth is lying hidden in the large number of unexplored plants in nature. These need to be explored before their extinction from the earth. The chemical compounds of the plants are also of great medicinal value and used for the various human and animal ailments. A large number of compounds belonging to different classes of compounds have been discovered and these have been compiled in the different dictionaries (Tables 1.1 and 1.2). Three research journals, namely, Phytochemistry, Journal of Natural Products, and Natural Products Communications and other journals are doing a great job by publishing latest research findings in the field of chemistry of natural products. The author of this book has 38 years of research experience in the field of chemistry of natural products and carried out research work on the chemical examination of a number of plant species. Chapters 5 to 7 of the book describe the author's research experience either alone or as a project leader or member of the research team in the isolation and structure elucidation of the chemical constituents, essential oils and fatty oils from the different parts of plants species.

REFERENCES

Aiyar, V.N., Dayal, R., and Seshadri, T.R. (1974). Triterpenoids of *Dalbergia sericea* (bark). *Curr. Sci.*, 43: 75.

Akhila, A and Rani, K. (1999). *Chemistry of Neem Tree* (*Azadirachta indica* A. Juss.). Progress in Chemistry of Organic Natural Products, Book Series, Springer-Verlag, Wein, Vol. 78, pp. 47–149.

Anon. (2009). *Training Module on Capacity Building of Communities Involved in Sustainable Forest Management*. Published by Forest Research Institute, Dehradun, India. pp. 123-128 (Katha), pp. 129–143 Natural Dye, pp. 145–150 (Tannins).

Bailey, J.A. and Mansfield, J. (1982). *Phytoalexins*. Blackie, Glasgow.

Bate-Smith and Swain (1962). Flavonoid compounds. In: Florkin M., Mason H.S. *Comparative Biochemistry III*. New York: Academic Press, pp. 75-80.

Bawa, R., Gupta, R, Rai, K., Dutt, B., Brahmi, M.K. [Eds.] (2009). *Advance Training on Non–timber Forest Produce*. Published by Dr. Rajan Bawa, Coordinator, Regional Centre, Nature Afforestation and Eco Development Board, Dr. Y.S. Parmar University of Horticulture and Forestry, Nauni, Solan, Himachal Pradesh, India. pp. 180–184.

Burkhill, I.H. (1935). *A Dictionary of Economic Products of the Malay Peninsula*. Published by the Crown Agents for the Colonies London, Vol. 2, 2402.

Camps, F. (1991). *Ecological Chemistry and Biochemistry of Plant Terpenoids*, J.B. Harborne and F.A.T. Barberan (Eds.) (Clarendon Press, Oxford) p. 331.

Culberson, C.F. (1969). *Chemical and Botanical Guide to Lichens Products*, University of North Carolina Press, Chapel Hill.

Dayal, R. (1986). Studies of Oleo-resins of *Pinus merkusii* and *Pinus insularis* from Arunachal Pradesh. *Ind. For.*, 112: 734.

Dayal, R. (1987). Phytochemical investigation of *Vateria indica* (leaves and roots). *J. Indian Chem. Soc.*, 64: 259.

Dayal, R. (1987). Occurrence of ursolic acid and related compounds in *Eucalyptus* hybrid leaves. *Curr. Sci.*, 56: 670

Dayal, R. (2015). *Natural Dyes.* I.K. International Publishing House Pvt. Ltd. New Delhi.

Dayal, R., Precha, V. Bartley, J. P. (1997). GC-MS studies of oil from *Eucalyptus* hybrid leaves. *Indian Perfumer*: 41(4): 60.

Dayal, R. and Dobhal, P. C. (2001). *Tetrameles nudiflora*- A new potential source of bryonolic acid. *Journal of Medicinal and Aromatic Plant Sciences*, 23 (4): 660–661.

Dewick, Paul M. (2009) *Medicinal Natural Products: A Biosynthetic Approach.* 3rd Edn. John Wiley and Sons.

FAO (2006). *Global Forest Resources Assessment* (2005) *Progress Towards Sustainable Forest Management.* FAO Forestry. Paper 147. Food and Agriculture Organization of the United Nations, Rome.

Finar, I.L. (1973 Reprinted 1996). *Organic Chemistry, Volume 1,* ELBS with Longman Singapore Publishers (Pte) Ltd. Singapore, pp. 698–731.

Finar, I.L. (1975). *Organic Chemistry, Volume 2,* ELBS with Longman Singapore Publishers (Pte) Ltd, Singapore, pp. 696–768.

Grayer, R.J. and Harborne, J.B. (1994). A survey of antifungal compounds from higher plants, 1982–1983. *Phytochemistry*, 37:19–42.

Harborne, J.B. (1993). *Introduction to Ecological Biochemistry,* 4th edn, (Academic Press London).

Hillis, W.E. (1986). Forever amber: A story of secondary wood components. *Wood Science Technology*, 203–228.

Howes, F.N. (1949). *Vegetable Gums and Resins.* Waltham, Mass Chronica Botanica Co. p. 188.

Kashiwada, Y., Nonaka, Goen-Ichiro., Nishioka, I., Chang, Jer-Jamg., Lee, Kyo-Hsiung (1992). Anti-tumor agents, 129. Tannins and related compounds as selective cytotoxic agents. *J. Nat. Prod.*, 55, (8)1033-1043.

Kaur, S., Dayal, R., Varshney, V.K., Bartley, J.P. (2001). GC-MS analysis of essential oils of heartwood and resin of *Shorea robusta. Planta Medica* 67: 883–886.

Kaur, S., Varshney, V.K., Dayal, R. (2003). GC–MS Analysis of essential oil of *Shorea robusta* Bast. *Journal of Asian Natural Products Research*, **5**(3): 231–234.

Khanbabaee, K., Ree, Teunis van. (2001). Tannins: Classification and definition. *Nat. Prod. Rep.*, 18, 641-649.

Langenheim, J.H. (1969). Amber: A botanical inquiry. *Science,* 163: 1157–1169.

Mendiratta, A., Dayal, R.; Bartley, J.P. (2005). GC-MS analysis of essential oils of needles and twigs of *Cephalotaxus harringtonia* var. *harringtonia, J. Essential Oil Research*, 17: 308–309.

Murthy, K.T. (1974). *Minor Forest Products of India: Their Availability and Demand, Utilization and Exploitation and Suggestions for Their Development.* Paper presented in Tenth Commonwealth Forestry Conference.

Shimizu, Y. (1986). *J. Amer. Chem. Soc.*, 108, 514.

Smith, J.E. and Moss, M.O. (1985). *Mycotoxins,* John Wiley, Chichester.

Tewari, D.N. (1994). *Tropical Forest Produce.* International Book Distributors, Dehra Dun, India, pp. 529-561.

Thomson, R.H. (1971). *Naturally Occurring Quinones.,* Second Edition. Academic Press, London, New York, pp. 1–38.

Walton, N.J. and Brown, D.E. (1999). *Chemical from Plants, Perspectives on Plant Secondary Products;* Imperial College Press and World Scientific Publishing Company Private Limited, pp. 1–25.

Würdig, G. and Woller, R. (1989). *Chemie des Weines,* Eugen Ulmer GmbH, Stuttgart.

2

CHAPTER

Essentials of Chromatography

The principle of chromatography and its applicability was discovered by a Russian botanist named Michael Tswett who described the technique for the first time in 1906 and published a detailed monograph in 1910. Tswett's observed the resolution of chlorophyll solution (in petroleum ether) on calcium carbonate stamped firmly into a narrow glass tube to various coloured zones which he termed as a chromatogram and the corresponding method, the chromatographic method. Chromatography uses the phenomena of adsorption. It may be defined as adhesion of a thin layer of a gas, a dissolved substance or a liquid to the surface of a solid body. An example of adsorption is the removal of coloured and other impurities from concentrated cane sugar solutions by charcoal. Tswett's chromatographic technique provided scientists with a particular efficient procedure for the preparation of chemical compounds in a high state of purity (Tswett, 1906). Isolation and characterisation of chemical substances, perquisites to examinations of composition and molecular structure, were brought to a new state of perfection. The findings began to make a wide use in 1930 and the chromatography was used by different researchers for the isolation of different compounds as given below (Kavarana, 1979).

S. No.	Researchers Name	Substance Separated
1.	Richard Khun and co-workers	Resolution of plant carotenes into several components
2.	Rimington and co-workers	Detailed technique for the separation of bile pigments porphyrins
3.	M.L. Wolform and co-workers	Separation of mixture of sugars on adsorption columns
4.	More and Stein	Separation of amino acids obtained by hydrolysis of proteins first on starch columns and later on ion exchange columns

TSWETT'S COLUMN

The most significant importance of Tswett's chromatogram is that it affords a very simple device for carrying out repeated adsorptions and elutions persistently. The chromatogram comprises essentially of a glass tube filled with a specially prepared adsorbent. The solution of the material to be separated is added to the top of the column of the adsorbent. The fresh solvent is percolated through the column after its drainage into the adsorbent. The granules of each layer of adsorbent and the small volume of the liquid that encloses them, act as a single batch adsorption stage. After the flow of solvent from the first layer of granules is complete, the solvent surrounds the granules of the next layer, a second batch of adsorption occurs, and a number of other batch of adsorption occur down the column to give equivalent of hundreds or thousands of individual stages. In this way a liquid flowing slowly through packed column achieves ceaselessly a number of operations which would require an impossible amount of time and labour to achieve singly. Thus, the adsorption and elution of the chemical compounds take place with the same solvent. Therefore, the molecules of a chemical compound in solution are persistently shuttling back and forth in a reversible equilibrium between the drifting solvent and the stationary adsorbent. As the proceeding solvent front carries these molecules lower to an unoccupied layer of adsorbent, the molecules depart the solution and are bound at the surface of the adsorbent. As soon as the influx of new solvent from above dilutes the dissolved material to less than a certain concentration, the adsorbed molecules are again discharged into solution. In this way a given compound travels down the chromatograph column as a zone, the leading edge of which is ceaselessly inching ahead, while the following edge is being gathered up. The relative affinity of a given compound towards the solvent and adsorbent determine the rate of migration of this zone. The great tendency of the given molecule to stick to the adsorbent will lead to slow movement of zone. In case of molecules having great tendency for the solvent than the zone will be moving rapidly. The degree of preference of a particular chemical compound for either the solvent or adsorbent can be determined quantitatively and assigned a numerical value known as distribution coefficient. The equal distribution of the chemical compound between solvent and adsorbent will lead distribution coefficient as 1 and resulting in a moderately fast moving zone. The distribution coefficient smaller than 1 will lead to a slow moving zone, while greater than 1 will result in a rapid moving zone. In fact, nature of adsorbent, solvent and chemical structure of the compound determines the distribution coefficient. The distribution coefficient of chemical compounds having varying structures for a given combination of solvent and adsorbent will be different (Kavarana, 1979).

CLASSIFICATION OF CHROMATOGRAPHIC METHODS

The chromatographic methods can be classified according to the physical nature of the mobile phase and stationary phase. The former (mobile phase) can be a gas or liquid, whereas later (stationary phase) can only be a liquid or a solid. Therefore, various combinations, such as liquid-liquid, liquid-solid, gas-liquid, and gas-solid are possible. When the separation of the compounds take place by a simple partitioning between two liquid phases, one stationary and one mobile, the process is called partition chromatography. If the physical surface forces are chiefly involved in the retentive ability of stationary phase, the process is called adsorption chromatography. These processes comprise liquid solution chromatography. When the mobile phase is a gas the methods are called gas-liquid chromatography and gas-solid chromatography.

Gas-liquid chromatography is very useful for the analysis of volatile compounds by offering high resolution, minimal analysis time, and sensitivity in parts per million range. The compounds which are not sufficiently volatile or thermally stable to be manipulated by gas chromatography can be analysed by liquid chromatography. The liquid chromatographic methods use a liquid mobile phase to transport the sample through the partitioning column which is packed with a solid adsorbent for a liquid coated on a solid. The mode of actions vary in two other chromatographic methods. In the ion-exchange chromatography the formation of true heteropolar chemical bonds takes place reversibly between ionic components in the mobile and stationary phases. The use of molecular sieves as the stationary phase brings about a categorization or of molecules based largely on molecular size. The method, actually exclusion chromatography, is called gel permeation chromatography by polymer chemists and as gel filtration by biochemists. The three major developments, namely, partition chromatography (paper and column), ion exchange chromatography and gas chromatography, were made after the original discovery of adsorption chromatography.

Partition chromatography constitutes the second major innovation in field of chromatography. Martin and Synge used silica gel packed columns as a support for stationary phase. Consden, Gordon and Martin introduced paper chromatography, a form of partition chromatography, in which strips or sheets of filter paper are used as a support for stationary phase. The two-dimensional systems described by them are useful, where one solvent system is inadequate for the resolution of all the compounds in a mixture. Normally, the stationary phase is aqueous, and the moving phase organic. Kritchevsky and Tiselius described the reversed phase paper chromatography for materials sparingly soluble in water. In this chromatography, the paper is impregnated with a hydrophilic substance and the aqueous phase becomes the moving phase. Due to its inherent simplicity and versatility, low equipment cost, and its ability to handle micro–quantities, paper chromatography became very famous quickly. It was used widely for the identification of natural products.

The ion-exchange phenomenon was first observed by two British agricultural chemists in 1850 who reported that soil can remove potassium or ammonium salts from water with release of an equivalent amount of calcium salt. Adams and Holmes in 1935 observed that certain synthetic resins were capable of exchanging ions, for example, sulphonic acid resins could be prepared as stable high capacity cation exchangers and polyamine-type resins exhibited anion exchange properties. Ion exchange chromatography finds use in the qualitative and the quantitative separation of both organic and inorganic ions. Separation of proteins and amino acids are the examples of organic ions, while fractionation of rare earth, and fission products are the examples of inorganic ions.

Gas chromatography was introduced by James and Martin in 1952 and this technique is now most widely used application of chromatography (James and Martin, 1952). The effect of chromatography was tremendous on all areas of analysis leading to the general progress of science. The impact was so great that Archer John Porter Martin and Richard Laurence Mellington Synge were jointly awarded Nobel Prize in chemistry for the year 1952 for their invention of partition chromatography. (Kavarana, 1979). Some of the later developments are High Performance Liquid Chromatography (HPLC), High Performance Thin Layer Chromatography (HPTLC), Gas Chromatography-Mass-Spectrometry (GC-MS) and Liquid Chromatography-Mass-Spectrometry (LC-MS).

CHROMATOGRAPHIC METHODS OF SEPARATION

Chromatography may be defined as the process of separation of a mixture of compounds by their distribution between two phases namely a fixed (the stationary phase) and the moving phase (the mobile phase). The different forms of chromatography are used for the separation of mixture of compounds and normally named by the nature of two phases used, the mobile phase is described first, e.g., liquid-solid, gas-liquid, etc.

The separation and purification of a mixture of compounds may be carried out using one or two or a combination of the following chromatographic techniques. The choice depends on the solubility properties and volatilities of compounds.

(a) Adsorption Chromatography
(b) Thin Layer Chromatography (TLC)
(c) Partition Chromatography
(d) Paper Chromatography (PC)
(e) Gas-liquid Chromatography (GLC)
(f) High Performance Liquid Chromatography (HPLC)
(g) Ion Exchange Chromatography

ADSORPTION CHROMATOGRAPHY

The process of separation of compounds from the mixture using a fixed solid phase and the moving mobile phase is generally called column chromatography. A glass column with a stopper fixed at the end of a suitable diameter is used for the stationary solid phase and a liquid is used as the mobile phase. The mixture of compounds may be dissolved in a suitable solvent and adsorbed on the solid, for example, silica gel and put on the top of the column. The pure solvent or a mixture of solvents may be used as a mobile phase for the elution of the column. The less adsorbed compounds on stationary phase are eluted first followed by strongly adsorbed compounds. The elution of the column with polar mobile phase will elute the column collectively and it will not be possible to save the compounds (Finar, 1975, reprinted 1996). The order of decreasing activity of some of the adsorbents is: activated charcoal >alumina >silica gel >magnesia >calcium carbonate >sucrose >starch >cellulose. The order of decreasing polarity of mobile phases is: Acetic acid >water >methanol >ethanol >propanol >acetone >ethyl acetate>ether > chloroform >methylene chloride >benzene >carbon tetrachloride >cyelohexane >petroleum ether.

THIN LAYER CHROMATOGRAPHY

It is a special type of adsorption chromatography in which a glass plate coated with a thin and uniform layer of adsorbent (silica gel, alumina, etc.) is used. The spots containing the small amount of mixture to be examined are put at the bottom of the plate and then placed vertically in the suitable developing solvent closed tank of proper size for the development of the chromatogram. The coloured components of the mixture can be seen readily with naked eyes, while the colourless compounds are observed either by placing in iodine chamber or spraying normally with 5 per cent sulphuric acid followed by heating in an oven at about 90 to 110°C till appearance of colour. The phenolic compounds can be detected by spraying with alcoholic ferric chloride and alkaloids with Dragendorff's reagent and other reagents suitable for other class of compounds. The spots can also be located by observing the plate in UV light. TLC method is used in separating all lipid compounds, i.e., the lipids, steroids, carotenoids, quinones, triterpenoids, flavonoids and their glycosides. TLC is preferred over PC due its versatility, speed and sensitivity (Finar, 1975). Silica gel as adsorbent on glass plates is used widely. The other adsorbents which may be used are aluminium oxide, celite, calcium hydroxide, ion exchange resin, polyamide, sephadex, cellulose, etc. Calcium sulphate hemihydrate (15%) is used as binder in the silica gel in order to stick it to the glass plates. The air dried plates are activated at about 110°C for 30 minutes. Silver nitrate is used for the argentation of TLC plates. The most recent type of TLC plates are coated with same fine microparticles of silica gel that are used in the columns of HPLC. This type of chromatography is termed High Performance Thin Layer Chromatography (HPTLC) and provides more efficient and rapid separation and finger printing of the crude extracts. TLC is generally carried out by ascent in the TLC jars saturated with solvent phase. The thickness (0.10–0.25 mm) is used for normal spot detection and 1 mm for preparative TLC.

HPTLC is advanced version of thin layer chromatography and is based on the adsorbent particles which are smaller than those used in conventional technique. These fine particles allow very good contact to be made between the sample, the mobile phase and the adsorbent surface. In normal TLC, 0.25 mm to 1.0 mm thick layer is used, while 0.1 mm thick for HPTLC. The improved efficiency and separation can be achieved by 3 cm to 6 cm run in comparison to 15 cm to 20 cm in conventional TLC. Thus, separation can be completed in 1/5 or 1/10 of time needed for conventional TLC. Details about this separation technique are available in a number of books.

PARTITION CHROMATOGRAPHY

In this type of chromatography, the stationary phase consists of a liquid substance strongly adsorbed on a solid column (as support) for example silica gel. The mixture of compounds to be separated is dissolved in a solvent (mobile phase) which is not miscible with the adsorbed solvent (the stationary phase) and the solution obtained is allowed to pass slowly down the column and subsequently followed by pure solvent.

The distribution of the compounds takes place between the stationary and mobile phases resulting in separation of compounds on elution depending upon their different partition coefficients. In normal practice, the stationary phase is generally water, and the water-immiscible solvent (or mixture of solvents) as the mobile phase (Finar, 1975).

PAPER CHROMATOGRAPHY

This is a special case of partition of chromatography in which the water adsorbed on a strip of paper acts as a stationary phase. The solution of very small amount of mixture to be analysed is spotted on a strip of paper leaving about 3 inches on one end of the paper. A number of spots can be put depending on the width of the paper and dried. Whatman paper (chromatographic grade) of the size 46 × 57 cm are generally used in paper chromatography.

The paper strip is placed in a suitable size of paper chromatography chamber saturated with mobile phase. The movement of the mobile phase can be ascending or descending order. Normally the later is preferred. The movement of the mobile phase phase is slow in paper chromatography. The movement of solvent front is marked and the paper is dried in the air. The coloured compounds can be observed with naked eye, while the colourless compounds are observed on spraying with a suitable spray reagent. For example, aniline hydrogen phthalate may be used for the detection of sugars. The authentic samples can also be spotted on paper for comparison of the R_F value. The ratio of distance travelled by a compound to the distance travelled by the solvent front is specific for each compound and is known as R_F value which depends on the experimental conditions (Finar, 1975). Two dimensional chromatography, provides a better method of separation than one dimensional method. Paper chromatography is used for separation and identification of water-soluble constituents, namely carbohydrates, amino acids, nucleic acid bases, organic acids and phenolic compounds.

GAS LIQUID CHROMATOGRAPHY

The gas liquid chromatography includes the separation techniques where the separation of the compounds is based upon the partition, or distribution of the analytes between the two phases in a dynamic system. In gas chromatography (GC), when the mobile phase is a gas and stationary phase as liquid, it is called Gas Liquid Chromatography (GLC) and with a solid stationary phase it is called Gas Solid Chromatography (GSC). The first research paper on gas chromatography was published by Martin and James in 1952. Perkin-Elmer, Hewlett-Packard, Griffin and George, Pye and others were first producers of commercial instruments with the packed column technique in the late 1950s and early 1960s. The chemically bonded fused silica capillary columns were introduced in the 1980s there by increasing the usefulness of GLC to a great extent. The mixtures of compounds to be analysed by GLC must be sufficiently volatile so as to present in the gas phase in experimental conditions for their transportation through the column. There is normally little association between the vaporised molecules of the compound to be analysed and those of carrier gas which streamlines the chromatographic process (Pattison, 1969; Fowlis, 1995).

The basic principle of gas chromatography is that greater affinity of the compound for the stationary phase will lead to longer retention of the compound before its elution and detection. The elution of the compound from the stationary phase depends upon its retention. The components of a gas chromatograph are given below.

(i) Column

The column is considered to be the heart of the gas chromatograph as the separation of the compounds takes place in it. The column is placed in a thermostatically controlled oven as the temperature will influence the volatility of the analytes. The packed column used for the analysis was later replaced by capillary columns in 1980s. The gas chromatographic packed columns consist of a suitable tube containing an inert solid phase (support) coated with a relatively non-volatile liquid phase which acts as a stationary phase. The separating power of the packed columns may be a few thousand plates for one meter long column. The capillary columns used in gas chromatography possess 0.1–0.5 mm internal diameter, 10–60 m length with the stationary phase (without solid support) is directly coated onto the column wall. The capillary column usually generate 50,000–1,00,000 or even more theoretical plates. It has been observed that the resolution is always not related to theoretical plates.

The stationary phase in the latest capillary columns is chemically bonded to silica column wall resulting in the extended temperature range of gas chromatography leading to better separation of the mixture. In addition the polar stationary phases can also be chemically bonded onto the column wall thus providing more degree of selectivity leading to the enhancement of the resolving power of theoretical plates (Fowlis, 1995).

(ii) Carrier Gases

The selection of carrier gas for gas chromatography takes into consideration its inertness, dryness, freedom from oxygen, safety, cost and availability. Nitrogen, helium and hydrogen may be used as a mobile phase in the gas chromatography. The use of hydrogen is avoided as there may be explosion in case of its leakage. Argon was largely used in argon ionisation detector and flame ion detector due its purity and availability. Majority of the users prefer helium as carrier gas for both GC-FID and for GC-MS. The carrier gas, consisting of a mobile phase is passed through column and the difference in the partition coefficient of individual compounds in the mixture between gaseous and stationary phases causes a separation of the mixture of compounds injected (Fowlis, 1995).

(iii) Injection System

The injection system provides a means of putting the sample or solution of the sample onto the column. A simple septum injector will be fine for the packed column but different injection systems are used for capillary columns. The main capillary injection devices are the split/splitless injector and the on-column injection. The injection device is helpful for maintaining the pressure and flow through the analytical column during the injection process and also prevents the ingress of air onto the column system. The injected sample is vaporised in a heated chamber and appropriate amount is allowed to pass into the column along with carrier gas.

(iv) Detectors

The suitable and efficient detector is required for the detection of individual compounds from the mixture eluted from the column. The selection of detector depends of the type of sample to be analysed and some of the detector used are.

(a) Flame Ionisation Detector (FID)
(b) Electron Capture Detector (ECD)
(c) Thermal Conductivity Detector (TCD)*
(d) Mass Spectrometer (MS)*
(e) Infrared Spectrophotometer (IR)*

FLAME IONISATION DETECTOR

It consists of a base in which the column eluent is mixed with hydrogen, a polarised jet and a cylindrical electrode arranged concentric with the flame. The air supply to the detector helps in combustion. The detector body is made of stainless or aluminium to which are fitted a flame ionisation coil and electrical connections to the collecting electrode, and a polarising voltage to the detector jet. The jet itself is usually insulated from the body of the detector. When the carrier gas and hydrogen are present only in the detector, the base current is very low. But on elution of the compound from the column and its presence and burning in the flame generate ions which move to the collector electrode due to the potential difference between the jet and the electrode. The resulting ion current is amplified and fed to the potentiometric recorder or data system. This detector is very useful for the organic compound detection and carbon less compounds will not be detected by it. FID is also insensitive to water and does not detect CO, CO_2, HCN and HCOOH. The wide linear dynamic range of the FID (~10^7) makes it very suitable for accurate quantitative analysis (Fowlis, 1995).

ELECTRON CAPTURE DETECTOR (ECD)

The ECD is a selective device invented for gas chromatography and is most sensitive for compounds containing electronegative elements, such as halogens. The detailed description about these and gas liquid chromatography have been given in a number of books.

---

* The brief description about first two detectors (a) and (b) is being described above. The details about all detectors have been given in a number of books.

HIGH PERFORMANCE LIQUID CHROMATOGRAPHY

High Performance Liquid Chromatography (HPLC), a powerful separation technique, was introduced by Huber and Hulsman in 1967. HPLC, also known as high pressure liquid chromatography, is a highly developed form of partition chromatography, a separation technique based on differential distribution of the compounds of a mixture between two phases of which one is stationary and the other mobile. The mobile phase in HPLC is a liquid while the stationary phase may be a solid (adsorption chromatography) or a liquid sustained on a suitable backing (liquid-liquid chromatography). These solid backup could be in the form of glass beds coated with a thin layer of silica microspheres (or some other adsorbents) or chemically bonded stationary phase in which the nature of adsorbent is modified by bonding organic groups to the adsorbent surface. The main components of the instrument include a solvent reservoir, a filter, a pump, a column, a detector and a recorder. The eluent from the solvent reservoir after filtration is pressurised and pumped through the chromatographic column. The mixture of compounds injected at top of column is fractionated into components, monitored by detector and recorded as peaks on the chart recorder. The method is specifically useful for the analysis of mixture of thermally labile or highly polar compounds which are difficult to be handled by GLC. The analysis time depends on the complexity of the mixture and the detection limit is usually in the range of 10^{-3} to 10^{-9}g. The reversed phase chromatography is the most popular form of HPLC because it can handle different variety of sample types. Reverse phase is opposite to the normal phase and in this the stationary phase is non-polar and mobile phase is polar (Snyder and Kirkland, 1979; Lindsay, 1992).

The HPLC equipment uses column filled with a solid stationary phase through which a liquid is pumped at high pressure (up to 12,000 p.s.i.). The liquids (mobile phase) including water, methanol, acetonitrile, dichloromethane, hexane or combination of some of these, are held in reservoirs. These liquids are drawn out of the reservoirs at precisely controlled flow rates, mixed if required, and thereafter pumped to an injector. The sample to be analysed is injected (10 μl) into the moving liquid stream, which moves onto to the top of the column. As the mobile phase elutes the sample components down the column, they separate, leave the column and enter the detector cell. The optical property of the mobile phase, viz., ultraviolet–visible absorption or refractive index is constantly checked, allowing the presence and concentration of each sample to be determined. The electrical information from the detector is amplified and transmitted to a recorder where a chromatogram is obtained as a visible record of the progress of each sample through the column. The performance of the pump, injector, detector and recorder can be completely automated.

The separation of the mixture components depends on the size, charge, adsorbing power or distribution coefficient. The column packing material consists of microscopic (10 μm diameter) silica sponges, whose surface is coated with a monomolecular layer of hydrocarbons material covalently bonded to silicon atoms. The mobile phase consists of chromatographic grade, water and methanol in varying proportions. Such a system is called a reversed phase chromatography, and separation occurs due to different tendencies of the components to distribute themselves between polar water/ methanol phase and non-polar hydrocarbon phase. In general, non-polar components spend more time in the stationary phase and elute after polar components.

The same parameters that ascertain the success of gas chromatographic separation are also applicable to liquid chromatography. Some of these are physical dimensions of the column, column packing, column temperature, liquid flow rates, sample size and injection technique. The column packing used in the reversed-phase is done by using a bonded phase material in which C-18 are hydrocarbon chains chemically bonded to the surface of silica particles of 10 μm diameter. These types are called ODS (octadecylsilane) bonded phases. The composition and flow rate of mobile phase (methanol or methanol-water) is controlled by microprocessor. Normally, the column length is 10–25 cm with internal diameter of 4.6 mm. The mobile phase is normally pumped through the column at a flow rate of 1–5 ml/min. If the elution of the column is done with one pure solvent the method is called isocratic elution. The technique is called gradient elution when the composition of mobile phase can be made to change in predetermined way during the separation. The gradient elution is used for a mixture of polar and non-polar compounds (Lindsay, 1992).

Advantages of HPLC over GC

(a) It is possible to isolate the compound in pure state.

(b) HPLC operation can be carried out at ambient temperature so that compounds are not subjected to thermal rearrangement.

(c) HPLC is mainly used for non-volatile compounds, e.g., higher terpenoids, alkaloids, phenolics of all types, lipids and sugars.

(d) Protein has been separated by HPLC on columns of modified sephadex, silica gels or ion exchangers.

(e) HPLC is very useful for clinical laboratory as the majority of compounds to be analysed are aqueous soluble, non-volatile and usually thermally unstable.

(f) Detectors choice is wide.

(g) The sample interaction with both the stationary phase and mobile phase is possible in HPLC. Complexation or hydrogen bonding can occur in HPLC mobile phase, but are absent in GC mobile phase. It is possible to increase the variety of stationary phases by suitable chemical modification of the silica surface. Due to all this HPLC is a more versatile method than gas chromatography (GC) and thus able to achieve much more difficult separations.

ION EXCHANGE CHROMATOGRAPHY

Two British agricultural chemists observed and studied the phenomenon of ion exchange in 1850 (Kavarana, 1979). The fundamental work on chromatographic separation by the use of ion exchange resins is mainly derived from the studies carried out in connection with the work on atomic energy USA. The main advantage of ion exchange chromatography is that it permits simple and rapid separation or concentration which is normally difficult and time taking. In ion exchange chromatography, the solid consists of a synthetic resin. Two types of resins namely cation exchanger and anion exchanger are used in this chromatography. Thus, it is possible to separate the mixture of bases and acids. The protein on hydrolysis gives a mixture of amino acids as- R.CH NH_3^+COOH]Cl^-. The amino acids will be retained in the column on passing this solution through a cation exchange resin, for example, Res–SO_3^-]H^+. The separation of amino acids may be carried out by elution of the column with buffer solution of different pH. The more weekly amino acids will be removed from the column first and the increase of pH will remove the more basic amino acids (Finar, 1975).

The efforts have been made to describe the very basic information about the chromatography used in the separation of different types of compounds. A number of books are available which can be referred for the detail information about chromatography.

REFERENCES

Finar, I.L. (1975). *Organic Chemistry, Volume 2,* ELBS with Longman Singapore Publishers (Pte) Ltd, Singapore, pp. 64–68.

Fowlis, Ian A. (1995). *Gas Chromatography,* John Wiley & Sons, Chichester, New York, Brisbane, Toronto, Singapore.

James, A.T. and Martin, A.J.P. (1952) Biochem. J., 50, 679.

Kavarana, H.H. (1979). *Ion Exchange Chromatography, Part* 1, Volume X, Number 2, March-April Published on behalf of Sarabhai M Chemicals, Baroda; Printed by Paper & Allied Products, GIDC, Industrial Estate Makarpura, Baroda.

Lindsay, Sandie (Ed. John Barnes). (1992). *High Performance Liquid Chromatography,* John Wiley & Sons Chichester, New York, Brisbane, Toronto, Singapore.

Pattison, J.B. (1969). A *Programmed Introduction to Gas-Liquid* chromatography, Heyden & Son Ltd, London.

Snyder, L.R. and Kirkland, J.J. (1979). *Introduction to Modern Liquid Chromatography,* John Wiley & Sons, Inc. New York, Chichester, Brisbane, Toronto, Singapore.

Tswett, M. (1906). Adsorption analysis and chromatographic method, Application to the chemistry of chlorophyll. Berichte der Deutschen botamischen (Gesellschaft 24, 384-393.

3

CHAPTER

Introduction to Spectroscopy

ABSORPTION SPECTRA

The molecule on absorption of light (electromagnetic wave of any wavelength), undergoes transition from the lower energy state to higher energy state. The energy absorbed is used to raise the energy levels of electrons in case of monoatomic molecules. The light absorbed by the diatomic molecule and containing more atoms-may cause changes in electronic, rotational and vibrational energy. The electronic transition give absorption (or emission) in ultraviolet and visible parts of the spectrum, whereas rotational and vibrational energy changes give absorption (or emission) respectively in the far and near infrared. The rotational and vibrational transitions may also occur with electronic transitions (Finar, 1975, reprinted 1996). The information on the structure of molecules is provided by the study of these energy changes. A schematic absorption spectrum is a plot of absorption of energy (radiation) against its wavelength (λ) or frequency (γ). An absorption may be characterised by two parameters, namely, the wavelength at which maximum absorption occurs and the intensity of absorption at this wavelength compared to base line.

A spectrum consists of distinct bands or transitions because the absorption (or emission) of energy is quantised. The energy gap of a transition is the molecular property and characteristic of molecular structure. The y-axis measures the intensity of the absorption band which is proportional to number of molecules observed (the Beer-Lambert Law) and to the probability of transition between energy levels. The absorption intensity is also a molecular property and both the frequency and intensity of a transition can provide structural information.

An absorption spectrum is obtained by placing the substance between the spectrometer and some source of energy that provides electromagnetic radiation in the frequency range being studied. The spectrometer analyzes the transmitted energy relative to the incident energy for a given frequency. The spectroscopic studies may be of two types. The electromagnetic radiation consists of waves composed of an electric vector which is perpendicular to the magnetic vector, and both vectors are perpendicular to the direction of propagation. The electric vector is mainly responsible for the light absorption by its interaction with the charged electron and atomic nuclei. Absorption spectra belonging to this type are UV, Visible, IR, Raman and Microwave spectra. The absorption spectra may also be obtained by interaction of charged particles with varying magnetic or electric fields. Nuclear Magnetic Resonance (NMR), and electron spin resonance belong to this type of absorption spectra (Finar 1975, reprinted 1996).

INFRARED SPECTROSCOPY

A person feels warm by sitting in front of fire. This happens due to radiant heat coming from the fire which was examined by Sir William Herschel in 1800. A solar spectrum of the sunrays was produced by putting a glass prism in the path of sunlight. The rise in temperature of a sensitive thermometer was observed when its bulb was moved from blue to red colour wavelengths and even beyond the red end of visible spectrum. The successive experiment indicated the portion beyond the red constituted a similar type of radiation to visible light as it could be refracted, reflected and absorbed by materials refracting, reflecting and absorbing visible light. The portion of a spectrum beyond red end of visible spectrum where this radiation is found is known as infrared and cannot be seen with the eye. Further researches have led to the discovery of many kinds of radiation, e.g., gamma rays, X-rays, ultraviolet (UV), visible, infrared (IR), microwaves and radio waves. These radiations travel at the same speed and vary only by their wavelengths. The visible and infrared regions constitute merely a small portion of electromagnetic spectrum. The visible red colour is nearest to the infrared in the electromagnetic spectrum (Cook and Jones, 1972).

Due to large variation of the wavelength in the different regions of electromagnetic spectrum it is generally denoted by using different terms, namely, nanometre (nm), micrometre (μm) and centimetre (cm). These units can be inter converted as given below:

$$1 \text{ m} = 10^2 \text{ cm} \quad (1)$$

$$1 \text{ m} = 10^6 \text{ μm} \quad (2)$$

$$1 \text{ m} = 10^9 \text{ nm} \quad (3)$$

The following inter conversions can be made on the basis of Eqns. 1 to 3

(a) By using Eqns. 1 and 2

$$10^2 \text{ cm} = 10^6 \text{ μm}$$

or

$$1 \text{ cm} = 10^4 \text{ μm}$$

$$\text{A cm} = \text{A} \times 10^4 \text{ μm}$$

(b) By using Eqns. 2 and 3

$$10^6 \text{ μm} = 10^9 \text{ nm}$$

or

$$1 \text{ μm} = 10^3 \text{ nm}$$

or

$$\text{B μm} = \text{B} \times 10^3 \text{ nm}$$

(c) By using Eqns. 3 and 2

$$10^9 \text{ nm} = 10^6 \text{ μm}$$

or
$$1 \text{ nm} = 10^{-3} \text{ µm}$$
$$C \text{ nm} = C \times 10^{-3} \text{ µm}$$

The infrared region of the electromagnetic spectrum includes the wavelength between 0.7 µm and 100 µm. The small portion of this region bound by wavelengths 2.5 µm and 15 µm (4000–666 cm^{-1}) is very useful. The region between 1400–650 cm^{-1} constitutes the fingerprint region and is the characteristic of a compound. The infrared wavelengths are always denoted in micrometres (µm). The positions of absorptions band in the infrared spectrum is described by wavenumber of the absorption. The wave number may be defined as the number of electromagnetic waves of wavelength λ which will occur in 1 cm.

We know that 1 cm = 1×10^4 µm

Thus, the number of waves in 1 cm can be determined by dividing the length (1 cm) by wavelength of radiation.

$$\text{Number of waves in 1 cm} = \frac{1 \text{ cm}}{\text{Wavelength}(\lambda) \text{ of radiation}} = \frac{1 \times 10^4}{\lambda}$$

1 cm has to be written as 1×10^4 µm due to expression of wavelength λ in micrometre. The wavenumber is written as cm^{-1} due to number of waves per centimetre.

For example, $$5 \text{ µm} \equiv \frac{1}{5} \times 10^4 \text{ cm}^{-1} \equiv 2000 \text{ cm}^{-1}$$

The above expressions indicate that the reciprocal of wave length multiplied by 1 cm expressed in micrometres is equal to the wave number, in cm^{-1} and is equal to $\frac{1}{\lambda} \times 10^4$ (Cook and Jones, 1972).

The radiation energy in the infrared (IR) region is not adequate to cause electronic excitation in most of the organic compounds, but capable of causing stretching and bending of organic bonds. The bonds in organic compounds can stretch and bend, and also absorb infrared radiation resulting to an IR spectrum showing a large number of peaks and a number of them overlap each other. The IR–spectrum is a plot of wavelength or frequency versus absorption recorded for a compound using double beam infrared spectrometer. The perpendicular scale of an infrared spectrum exhibits either percent transmission (%T) or absorbance (A) of the radiation passing through the sample.

$$(\%T) = \frac{\text{Intensity}}{\text{Original intensity}} \times 100 \qquad A = \log \frac{\text{Original intensity}}{\text{Intensity}}$$

The base line in the usual infrared spectrum is exhibited at the top of the paper representing no absorption of energy by the sample. When the sample absorbs radiation at a wavelength, the tracing indicates a decrease in %T or an increase in absorbance. The absorption is thus recorded as a dip known as an absorption peak or absorption band;

the wavelength or frequency corresponding to the minimum point of an absorption band is used to identify that band.

The vibrating atoms absorbs energy to give an absorption band in the infrared spectrum. Let us take that C atom is held in a fixed position and a spring is attached to it at the end to which different atoms may be attached. The diatomic molecules containing carbon and hydrogen will vibrate at a particular frequency on stretching and releasing. The replacement of H by C at the end of spring will lead to vibration of the spring at a different slower frequency due to higher carbon mass and still slower on replacement of carbon with iodine at the end of the spring. Thus, the frequency of vibration is inversely proportional to the vibrating masses. Therefore, the absorption band for these diatomic atoms will occur as given below.

Diatomic atoms	Position of absorption band
C–H	3000 cm^{-1}
C–C	1000 cm^{-1}
C–I	500 cm^{-1}

If an additional spring is attached to the fixed carbon leading to the two ways of stretching in each molecule resulting in the two absorption bands (H–C–C, 3000 cm^{-1} and 1000 cm^{-1}) Two absorption will also appear for the two identical bonds together (C–C–C) due to symmetric (in-phase) and asymmetric (out-of-phase) vibrational motion. The increase of bonds between two atoms say from single to triple (C–N to C≡N) increases the frequency of vibration due to strength of the bond between two atoms [1070 cm^{-1} (C–N), 1650 (C=N), 2250 cm^{-1} (C≡N)]. Six vibrations of the group namely, asymmetric and symmetric stretches, in-and out of plane symmetric and asymmetric deformations are observed. Three types of absorption bands, namely, fundamental, combination and overtone are observed in the creation of infrared spectrum. The later two bands are weaker than the first one. Vibrations different to stretching of a bond can also occur in the molecule. These are known as deformations and refer to the changing of bond angle between the atoms of a molecule and occur at frequencies lower than those of stretching vibrations. The four types of deformations, namely, scissors, rock, wag and twist are the general deformational vibrations applicable to –(AB_2)– groups of atoms (Cook and Jones, 1972).

It is difficult to interpret the IR spectrum completely. The high frequency end above 1500 cm^{-1} (4000–1500 cm^{-1}) of the IR spectra is useful for the identification of functional groups due to relatively little overlapping of the peaks in this area. The absorption at ~3300 cm^{-1} and 1700 cm^{-1} indicate the presence of –OH (hydroxyl) peaks and >C=O (carbonyl) peaks respectively. The change in the dipole moment of a bond, as its length changes, is responsible for the intensity of an IR stretching absorption. The dipole moment of a bond can be defined as the product of the distance between two charges in a dipole and the magnitude of charges. Due to high polarity of carbonyl groups, changing the bond length markedly changes its dipole moments. The symmetrical bonds like double bond in ethylene or triple bond in acetylene having no dipole moments do not exhibit IR stretching absorptions. The IR spectrum with frequency below 1500 cm^{-1} exhibits a baffling pattern of overlapping stretching and bending absorptions. This region is often called finger print region due to appearance of so many peaks in this region. The finger print region is specific for a compound as it contains a great deal of information about the pattern of substitution in the structure of the compound instead of types of functional groups. The interpretation in this region is more difficult due to presence of so many different absorptions in comparison to higher–frequency region. The superimposable IR spectra of an unknown compound with a sample of a known compound can be used to identify it, since no two different compounds have absolutely identical IR spectrum.

Generally, the organic chemists are contented to detect the important functional groups or patterns of substitution on double bonds or aromatic rings from the spectra. The IR specialists can derive the detailed information about the structure of the compound. IR cells made of common salt (sodium chloride) transparent to most IR radiations are used for recording the IR spectrum instead of glass and quartz cells which absorb IR radiation. Carbon tetrachloride, carbon disulphide, and chloroform are generally used for recording the IR spectrum of the compound in a solution by using a double-beam spectrometer. The IR spectrum of the compounds which are insoluble in CS_2, CCl_4 may be recorded by grinding into fine powder along with dry potassium bromide and passing into pellets which is transparent to IR radiation. The mull of insoluble compound made with nujol is also used for recording the IR spectrum, specially for the detection of hydroxyl and carbonyl functional groups. Nujol absorption bands also appear in the IR spectrum which need to be ignored. The exhaustive description about the infrared spectroscopy is given in a number of books (Nakanishi and Solomon, 1977).

ULTRAVIOLET AND VISIBLE SPECTROSCOPY

The wavelength (λ) of the radiation consisting of the visible light varies from 400–800 nm. The energy of light radiation and its wavelength are inversely proportional to each other and the energy of photon of visible light have been calculated and varies from 72 kcal/mole (400 nm) to 36 kcal/mole (800 nm). The ultraviolet (UV) light can be divided into near UV and far UV depending the proximity of their wavelength to those of visible light. The near UV wavelength varies from about 200–400 nm, and easily identified by UV spectrometers. The near-UV light energy varies from 72 kcal/mole of wavelength 400 nm to 143 kcal/mole of wavelength 200 nm and is therefore greater than that of visible light. The wavelength in far-UV region spreads from 100–200 nm.

Ultraviolet spectroscopy usually refers to electronic transitions occurring in the region of the electromagnetic spectrum (wave length λ in the range 200–400 nm) accessible by UV spectrometers. The absorption in the visible range (400–750 nm) also takes place due to electronic transitions and are also accessible by the UV-VIS instrument, but less useful in the structure elucidation of organic compounds due the colourless nature of organic compounds (Finar, 1975, reprinted 1996).

The absorbance of radiation by the molecules will take place when its wavelength corresponds to the energy required to bring about some change in the molecule. The excitation of an electron from normal orbital into higher energy-orbital is the change that takes place in the ultraviolet and visible light and as a result of this, the molecule is said to have raised from its ground state to an excited state. A vast number of molecules, present in small quantity of chemical compounds, are also associated with slightly different ground states and excited states because they can occur in different conformations and their bonds may possess slightly different vibrational and rotational energies. Therefore, the compounds will absorb light over a range of wavelengths.

The total energy of a molecule is the sum of its electronic, vibrational and rotational energies. The electronic energy is more than vibrational which in turn is more than rotational energy. The energy absorbed in UV region produces changes in the electronic energy of the molecule resulting in transition of valence electrons in the molecule. The following transitions may take place and their general order of energy difference is (Finar, 1975, reprinted 1996):

$$\sigma \rightarrow \sigma^* > n \rightarrow \sigma^* > \pi \rightarrow \pi^* > n \rightarrow \pi^*$$

The correlation between the absorbed energy in an electronic transition and the frequency v or wavelength of radiation causing the transition is given by the Eqn. (1).

$$\Delta E = hv = \frac{hc}{\lambda} \tag{1}$$

where h is Planck constant and c is the velocity of light.

ΔE represents the energy difference in an electronic transition in a molecule from the low energy state (ground state) to high energy state (excited state). The amount of energy absorbed is related to energy difference between ground state and excited state. The smaller energy difference leads to longer wavelength of absorption. The position and intensity of absorption band is its main characteristics. The position of absorption corresponds to wavelength of radiation whose energy is equal to that required for an electronic transition. The intensity of absorption is largely dependent on two factors namely interaction probability between the radiation energy and electronic system for raising the ground level to an excited state and polarity of the excited state. A simple expression for the intensity can be derived from the Lambert-Beer law which states that proportion of light absorbed by a solute in a transparent solvent is independent of intensity of the incident light and proportional to the number of absorbing molecules in the light path.

$$\log_{10}\left(\frac{I_0}{I}\right) = A = \varepsilon cl$$

where I_0 = Intensity of incident light
I = Intensity of transmitted light
ε = Molar absorptivity or molar extinction coefficient or molecular absorption coefficient
c = Concentration of solute in moles/litre
l = Cell (path) length in cm
A = Absorbance

The UV spectrum is a plot of the wavelength or frequency of absorption against the absorbance $\log_{10}\left(\frac{I_0}{I}\right)$ or the transmittance $\frac{I}{I_0}$

The value $\log_{10}(I_0/I) = A$ is known as absorbance of solution and normally recorded directly by UV or visible light spectrometers for each wave length of light. The value of ε can be calculated easily for a compound at any wavelength.

The ability of a compound to absorb light at any given wavelength is measured by its molecular absorption coefficient, ε. The wavelength at which the molecule has its highest absorption coefficient is designated λ_{max} and the corresponding absorption coefficient as ε_{max}. The different maxima in a compound are associated with different characteristic values of ε_{max}. The ε_{max} greater than 10^4 and less than 10^3 corresponds to high intensity and low intensity absorption, respectively.

When the molecule absorbs light, an electron is excited from its ground state orbital to a higher level orbital and thus moved farther away from its best location near the atomic nuclei. The alkanes contain C–C and C–H σ bonds and thus all the electrons are tightly held by the molecule and therefore, high energy radiation is needed to excite these electrons. The alkanes thus absorb around 135 nm and difficult to be detected by UV spectrometers. It is easy to excite electrons in π-bonds, but unconjugated alkenes still absorb near 170–180 nm. But in the conjugated alkenes, the absorption shift to longer wavelength (lower energy) radiation. The reason being that the excitation of an electron from an orbital in a conjugated double bond leaves the remaining electrons free to spread out over all the p orbitals. The excited state in the conjugated system is easier to be formed as it is stabilised by resonance than the excited state from an unconjugated double bond. It is still easier to shift an electron into an upper energy level in the compounds having more than two double bonds in conjugation, and even longer wavelengths will provide the required energy. Due to this, the excitation of compounds with seven to eight double bonds is quite easy and as a result of this λ_{max} of these compounds lies in the visible region. It has been observed that λ_{max} for the conjugated compounds increases with the number of double bonds and also with the number of alkyl substituents on the double bonds.

It has been observed that almost all the carbonyl compounds absorbs light near 280 nm due to excitation of unshared electrons on oxygen. The conjugated carbonyls have comparatively intense absorption than the unconjugated carbonyl. The aromatic compounds and compounds with unsaturated hetrocyclic rings also absorb near–UV light and exhibit different maxima in their spectra. The strongly coloured compounds like carotenoids have a long sets of conjugated double bonds along with heteroatoms (normally nitrogen or oxygen) in conjugation with double bonds.

Chromophore: A covalently unsaturated group, namely, $>C=C<$, $>C=O$, NO_2, etc. responsible, for light (electronic) absorption in the ultraviolet region or colour in visible region are known as chromophore.

Auxochrome: The attachment of a substituted group like –OH, $-NH_2$, etc., to a chromophore, changes both the wavelength and intensity of absorption is known as auxochrome.

Bathochromic shift: The absorption shift to longer wavelength due to substitution or solvent effect is known as Bathochromic shift.

Hypsochromic shift: The absorption shift to a shorter wavelength due to substitution or solvent effect is known as Hypsochromic shift.

The absorption bands observed in the organic compounds containing some of the chromophores and auxochromes are given below (Table 3.1).

TABLE 3.1 Absorption bands in different compounds (Finar, 1975, reprinted 1996)

Compounds	Absorption region, nm
Alkanes With auxochrome	140–150 (σ–σ^*) 170–175 (n–σ^*)
RBr	200–210
RI	255–260
ROH	180–185
RNH_2	190–200 (n–σ^*, π–π^*, n–π^*)
Homoannular diene	253 (n–σ^*, π–π^*, n–π^*)
Heteroannular diene	214
Aldehydes	180 (π–π^*) 290–295 (n–π^*)
Ketones	190 (π–π^*) 270–280 (n–π^*)
Saturated monocarboxylic acid	200–210 (n–π^*)
Esters	200–205 (n–π^*)
Amide	205–220 (n–π^*)
Crotonaldehyde	220 (π–π^*) 321 (n–π^*)

The different bands in benzenoid compounds are sometimes referred to by letters, e.g., *E*(180–200), *K*(220–250), *B*(250–290), *R*(275–330). The π–π^* transitions give rise to *E*- and *B*-bands. The π–π^* transitions also give rise to K-band in the aromatic compounds with benzene ring directly attached to a group containing a multiple bond, e.g., C_6H_5CHO, $C_6H_5NO_2$, C_6H_5COOH and other compounds. If the group containing a lone pair of electrons, is directly attached to a benzene ring, then *n*–π^* transitions are possible giving rise to *R*-band which will be hidden by B band due to its greater intensity and overlapping with *R* band (Finar, 1975, reprinted 1996).

Ethanol is normally used as solvent for recording the UV spectrum of the organic compounds due to their solubility in it and transparency above 200 nm. The UV spectra are very useful in the structure elucidation of organic compounds isolated from different plant species specifically, the flavonoids. UV spectra play a very important role in assigning the position of phenolic hydroxyls in flavonoids thus providing a great help in their structure elucidation. The flavonoid spectrum is normally recorded in methanol or ethanol solution. The flavonoid spectrum typically consists of two absorption maxima occurring in the ranges 240–285 nm (band 2) and 300–550 nm (band 1). The variation can be caused by different hydroxylation pattern. The basic absorption bands observed in different types of flavonoids are as given below (Table 3.2).

Table 3.2 Absorption ranges of flavonoids in ultraviolet-visible region (Mabry et al., 1970; Harborne et al., 1975)

S. No.	Type of flavonoid	Band 1, nm	Band 2, nm
1.	Flavone	310–350	250–280
2.	Flavonols (3–OH substituted)	330–360	250–280
3.	Flavonols (3–OH absent)	350–385	250–280
4.	Isoflavones, Isoflavones (5–deoxy–6, 7–dioxygenated)	310–330 (shoulder) c.320 peak	245–275
5.	Flavanones and dihydroflavonols	300–330 (shoulder)	275–295
6.	Chalcones	340–390	230–270 (low intensity)
7.	Aurones	380–430	230–270 (low intensity
8.	Anthocyanidins and anthocyanins	465–560	270–280

The different shift reagents like aluminium chloride ($AlCl_3$), aluminium chloride ($AlCl_3$)+hydrochloric acid (HCl), sodium acetate (NaOAc), sodium acetate (NaOAc) +boric acid (H_3BO_3), sodium methoxide (NaOMe) are used for determining the position of hydroxyl groups in the flavonoids. The details have been described in standard reference books as given below.

1. Mabry, T.J., Markham, K.R. and Thomas, M.B. (1970). *The Systematic Identification of Flavonoids*. Springer, New York.
2. Harborne, J.B., Mabry, T.J. and Mabry, H. (1975). *The Flavonoids*. Chapman and Hall, London.
3. Harborne, J.B., and Mabry, T.J. (Eds.) (1982). *The Flavonoids: Advances in Research*. Chapman and Hall, London.

NUCLEAR MAGNETIC RESONANCE SPECTROSCOPY

Nuclear magnetic resonance (NMR) spectroscopy, specifically that of proton is one of the most effective techniques for the structure elucidation of the chemical compounds. The phenomenon of Nuclear Magnetic Resonance is restricted to chemical compounds comprising isotopes with either odd atomic number or odd mass number. Some of the examples comprise ^{1}H, ^{13}C, ^{19}F, ^{31}P, ^{15}N. The nuclei which are competent of resonance act as small magnets, when placed in an external field and can assume to $2I + 1$ orientations with respect to the direction of the field, where I denotes the spin quantum number of the nucleus. The spin quantum number (I) for the proton (^{1}H) is ½ and thus can assume two orientations in a magnetic field, which can be represented by alignment with field (low energy) and against the field (higher energy). The difference of energy between two states is related to the magnitude of the external magnetic field (H_0) and represented as:

$$\Delta E = 2\mu H_0 \quad (1)$$

where μ = The magnetic moment of the nuclear magnet and is constant for a particular nucleus.

Electromagnetic radiation is a form of energy whose magnitude, E, is given by the expression:

$$E = h\nu$$

where h = Planck's constant; ν = Frequency of radiation

The transition from the lower energy state to higher energy state may be supported by absorption of electromagnetic radiation having frequency ν by the nucleus when

$$\Delta E = h\nu \quad (2)$$

Taking Eqns. (1) and (2) into consideration

$$2\mu H_0 = h\nu$$

or $$H_0 = h\nu/2\mu$$

or $$\nu = 2\mu\, H_0/h$$

Thus, $$\nu \;\alpha\; H_0$$

The instruments use magnetic fields of such intensity, that radiation in the radio frequency range is needed to obtain the resonance condition. The nuclei in the excited state can come back to ground state by a process known as relaxation. The absorption band in the n.m.r is normally recorded by sweeping the magnetic field whilst irradiating the sample with radiation of constant frequency. The absorption of radiation takes place on reaching the field value of $h\nu/2\mu$ (Haws et al., 1973).

SHIELDING AND DESHIELDING

The hydrogen nuclei in organic compounds are circled by valence electrons. The movement of these electrons induced by an applied magnetic field, which contributes to a small rise with the secondary field in the proximity of protons. This induced field performs against the applied field and shields the protons from its impact. Thus, the proton can be brought into resonance by increasing the applied field to over this Diamagnetic Shielding effect. Therefore for resonance, field applied has to be increased as given below.

$$H = H_0 + H'$$

H_0 = Normal applied field

H' = Extra applied field to overcome shielding effect

The extent of shielding mainly depends upon the electron density in the proton region and as a result of which protons in various chemical surroundings are magnetically different and require varying field enhancement to prompt them into resonance. The three sets of protons in ethanol thus resonate at various field strengths.

The electron negativity of the neighbouring atoms, in an organic compound containing hydrogen atoms causes inductive effect which mainly affects the electron density in the proximity of the proton. Therefore, the protons under large inductive effect will require lower extra field to bring about the resonance. The methylene protons in the compound ICH_2CH_2Cl attached to more electronegative chlorine absorb at lower field than the ethylene protons attached to iodine. The induced magnetic fields due to multiple bonds separate the immediate environment into two regions, one in which the proton is shielded, the other in which it is deshielded with respect to applied field. For example, the alkynes protons are shielded in comparison to alkenes protons and thus absorbs at higher field (1.8 to 3.1 ppm) than those of alkenes (4.6 to 7.7 ppm).

The aldehydic proton, due to presence of carbonyl group is comparatively deshielded more and thus absorbs at lower fields value (9.3 to 10.0 ppm). In aromatic compounds, the π–electrons are specifically susceptible to an applied field and thus a comparatively high (in comparison to alkenes) induced field is set up. Therefore, aromatic proton in general and benzenoid protons (6.6 to 8.5 ppm) have more δ-value than alkenic protons (Haws et al., 1973).

CHEMICAL SHIFT

The absorption band positions in ^{1}HNMR spectrum is recorded quantitatively in comparison to a standard tetramethylsilane (TMS). The difference between the resonance position of a specific proton and that of TMS protons is known as chemical shift and is determined as given below.

$$\delta(\text{parts per million}) = \frac{\text{Frequency of TMS signal} - \text{Frequency of sample signal}}{\text{Spectrometer frequency}} \times 10^6$$

τ(tau-value) can be calculated as $10-\delta$.

Both δ and τ values do not depend on the instrument used, but possess a characteristic value for a specific type of proton. The chemical shift and chemical environment of the protons are closely related. The absorption band area is directly proportional to the number of protons accountable for the signal (or causing the signal).

SPIN-SPIN COUPLING

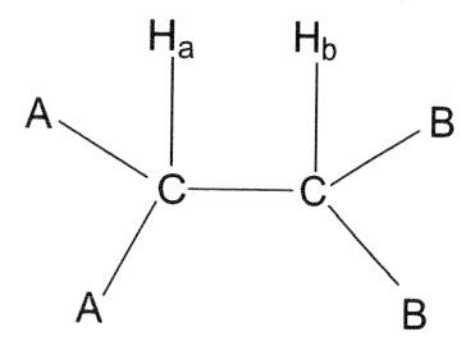

1

In a simple organic chemical compound (1), the exact amount of magnetic field required for the resonance of H_a at any one instant will depend on the orientation of H_b (Aligned with or against the field).

Therefore, the signal for H_a will appear as doublet and similarly that of H_b will also be a doublet. The separation between the two components of the doublet is known as coupling constant and denoted as J and measured in hertz. Taking the above reasoning for the compound 1 into consideration, the signal for H_a in compound 2 and compound 3 will appear as triplet and quartet, respectively. Normally, the splitting of a proton signal in the n.m.r. is equal to number of identical protons (n) to which it is coupled, plus one ($n + 1$).

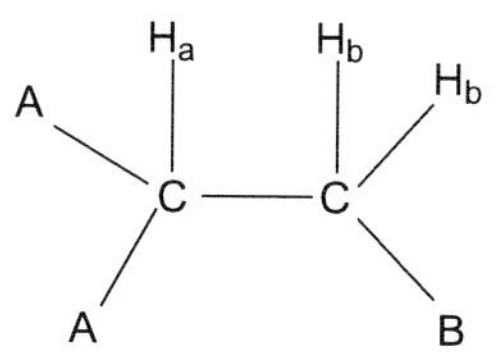

2

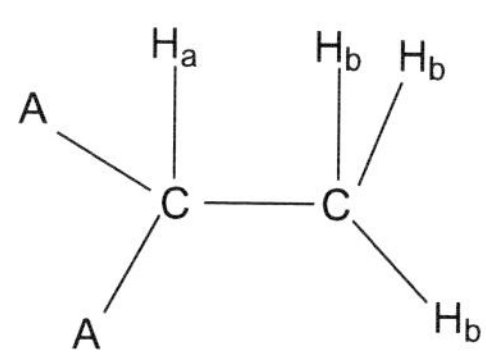

3

The alphabetical notation can also be used for describing the spin-spin coupling patterns of the protons. When the difference in chemical shift is large, the considerably separated alphabet will be assigned and the ones having similar chemical shift are assigned letters adjacent to one another in the alphabet. Some of the examples are as given below.

CH_3CHO	$C_2H_5CO–CH_2–CH_2–COCH_3$	$Cl.CH=CH_2$
AX_3	A_2B_2	ABC

The prime is used for the chemically, but not magnetically equivalent protons.

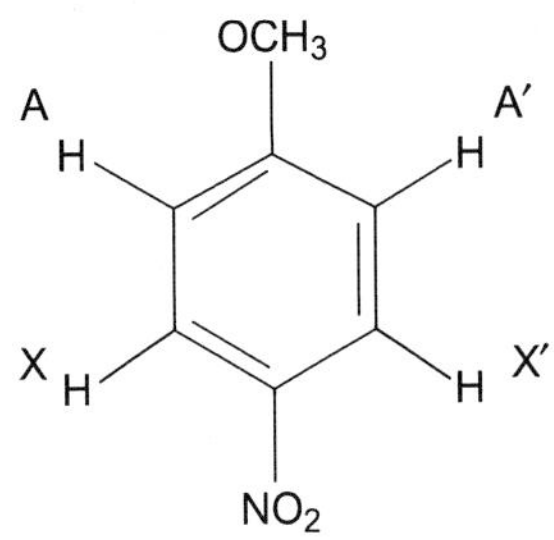

AA′XX′

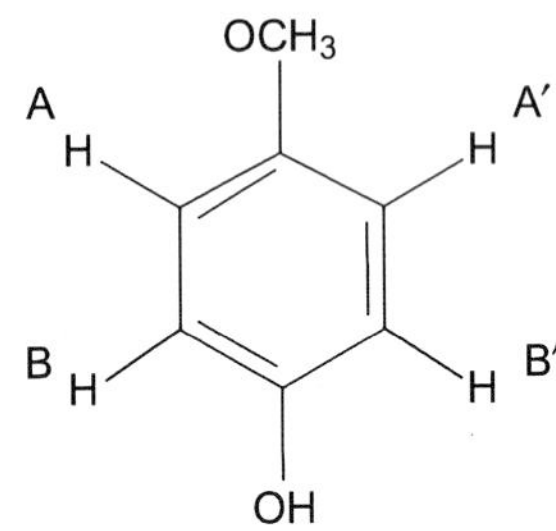

AA′BB′

The more complexity is observed in the coupling patterns on changing an (AX) system into an (AB) system. The spectra can become very mixed up when $\Delta\,\delta \leq J$ (Haws et al., 1973). Efforts have been made to describe a precise basic introduction about the ^{1}HNMR. Very considerable development have taken place in the field of NMR spectroscopy and have been described in various books.

FT NMR SPECTROSCOPY

The introduction of Fourier transformation (FT) technique in nuclear magnetic resonance (NMR) spectroscopy in 1966 by Ernst and Anderson started a new era in this branch of spectroscopy (Ernst and Anderson, 1966). The requirement for recording highly resolved spectra of small amount of pure compounds and less abundant nuclei in a short time has led faster growth of FT NMR spectroscopy. The NMR spectral parameters acquired using the FT technique consist of chemical shifts, spin couplings, quadrupolar coupling constants dipolar interactions and different types of relaxation times, such as spin-lattice (T_1), spin-spin (T_2) and the spin–lattice relaxation in the rotating frame. Structural and dynamical evidence made accessible through the data from FT-NMR has made this technique very useful. The broad features of developments in the area of FT-NMR research in the beginning were classified into four categories as given below (Narasimhan, P.T).

1. Polarisation the nuclear spin system
2. Excitation and coherence in the spin system
3. Altering the NMR Hamiltonian-Decoupling and selective averaging
4. Fourier Transformation

The details about Fourier Transformation (FT) have been described in a number of books and a brief description about FT is being given below.

Fourier Transformation (FT): The NMR signal obtained as a function of time, $f(t)$, can be Fourier transformed to provide NMR spectrum $F(w)$ in the frequency domain. The discrete FT version is used in computers using the Cooley-Tuhey algorithm. It is important to note that the pulse excitation of the spin system resulted in a response, viz., the FID. There is a cause and effect relation between excitation and response. The FT technique can be perceived as a procedure for obtaining a cross correlation between the excitation at time $t = 0$ and the effect following this excitation. The versatility of FT method can be extended by changing the form of excitation from pulse to other forms, such as stochastic excitation (Ernst, 1970), or rapid scan excitation (Dadok and Sprecher, 1974) or tailored excitation (Tomlinson and Hill, 1973) and so on and cross-correlating the response with appropriate excitation functions.

It is possible to equip the NMR system in a coherent state and allow it to develop in the presence of chosen interactions for a time t_1 and then find the signal after a period t_2. Fourier transformation of the signal can be done as a function of two time variables t_1 and t_2. The spectrum obtained is called 2-dimensional FT NMR spectrum or 2D spectrum. The selected interactions may take place during the mixing period t_m which exists between the evolution period t_1 and the detection period t_2. These processes can, in turn, be studied through appropriate 2D spectra. The idea of 2D spectroscopy was discovered by Jeener in 1971 and has been vastly developed by Ernst and co-workers into a potent tool (Jeener, 1971; Ernst et al., 1981). A very short introduction about FT-NMR techniques has been described above.

The FT NMR spectroscopy was found to be of immense importance and as a result of which Professor Richard R. Ernst of ETH, Zurich, Switzerland was awarded Nobel Prize in chemistry in 1991 for his contribution to the development of the methodology of high resolution nuclear magnetic resonance (NMR) spectroscopy.

Pulsed Fourier Transform NMR

The operation of modern NMR spectrometers takes place in a pulse mode in which the nuclei whose precession frequencies (chemical shifts) are to be perceived are subjected with a powerful radio-frequency pulse terminating only for a few microseconds. This very short pulse period causes the reorientation of all the nuclei in the magnetic field and leaving them in a state of non-equilibrium at the end of the pulse. In the halfway between pulses, the precessing of nuclei takes place at frequencies depending on chemical shifts and spin-spin couplings affecting the nuclei. Different relaxation processes causes the slow disappearance of the precessing magnetisation. (Shoolery, 1984).

The precession of each group of chemically equivalent nuclei generates an exponentially withering sinusoidal voltage in the receiver coil encompassing the sample. All these waveforms taken together forms a composite, time-dependent voltage named the "free induction decay" (FID) which is the sum of diminishing sine waves of different frequencies. It is possible to separate these frequencies with, Fourier transformation (FT) and indicate them as a "spectrum" of lines on a frequency scale, namely the NMR spectrum.

The split in time of the nuclear excitation and successive finding of the precession frequencies is a key benefit of pulsed FTNMR. The molecules of small and medium size containing both the protons and ^{13}C-nuclei have relaxation time of the order of seconds, leading to several interactions between nuclei before finding the NMR signals. This development will be helpful in finding out the number of protons bonded to each carbon atom in a molecule and also in additional particular observation of detailed interactions between magnetic nuclei. The details have been described (Shoolery, 1984).

The pulsed-Fourier Transform NMR spectrometer is not significantly different from the swept spectrometer. Its main components are as given below.

(i) A magnet, which may be permanent, an electromagnet or a superconducting solenoid.

(ii) Several radio frequency sources.

(iii) A detector

(iv) A minicomputer for carrying out both signal averaging and the Fourier Transform

The instrument applies a short burst of high powered radio frequency energy and then set in motion the computer to begin sampling the nuclear response from the detector at a relatively high ratio of speed (Cooper, 1978). One of the main advantages of the uses of FT in NMR has been the increased sensitivity made feasible by the multiplex benefit, along with the availability of higher magnetic fields. The insensitive nuclei like ^{13}C and ^{15}N having low natural abundance can now be studied as their weak signals cannot be observed with swept fields. Thus, the application of NMR is possible to many complex systems. The benefits of FT method have increased after the introduction of techniques which uses sequences of RF pulses of various amplitude and phase. The chemists can now find out the number of hydrogens bonded to each carbon in a molecule by using sequence methods conveying acronyms such as APT, DEPT, COSY and NOESY. It is also possible to predict the complex coupling patterns, and connectivity between bonded and non-bonded atoms. The availability of the Fourier Transform technique had made possible the development of completely new set of experiments by using pulse sequences (King and Williams, 1989).

King and Williams have published the detailed description about the Fourier Transform in chemistry about 26 years earlier (King and Williams, 1989; 1989a; 1990).; (Williams and King, 1990, 1990a). The NMR of a compound can be recorded as one- and two-dimensional NMR. Simple single-phase experiments and multiple-pulse experiments such as DEPT and the attached proton test (APT) are the examples of one-dimensional spectroscopy. The name one-dimensional points to experiments in which the transformed signal is exhibited as a function of a single frequency, e.g., the range of Larmor frequencies of either ^{1}H or the ^{13}C nuclei in the molecule. Similarly, the coordinate axes correspond to two frequency domains in a two-dimensional spectrum. In majority of cases, these are two chemical shift ranges, but sometimes one axis represents a range of coupling constants. The perceived signal which is a function of both parameters constitutes a third dimension. The information derived from various one- and two-dimensional NMR experiments have been very briefly described in chapter 4 of this book (Agrawal, 1994).

The use of Fourier Transform methods has revitalised familiar analytical techniques like NMR and IR spectroscopy. The glossary of some of the special terms associated with FT-NMR spectroscopy have been described (King and Williams, 1990). It is now possible to elucidate the structure of complex naturally occurring compounds with this sophisticated technique and IR, UV, MS spectra with meticulous detail.

The detailed information about the NMR can be obtained from a number of books published on the subject. The titles of some of the books are given below.

1. Derome, A.E. (1987). *Modern NMR Technique for Chemical Research*, Pergamon Press, New York.
2. Ernst, R.R., Bodenhausen, G. and Wokaun, A. (1987). *Principles of Nuclear Magnetic Resonance one and two Dimensions*, Clarendon Press, Oxford.
3. Sanders, J.K.M. and Hunter, B.K. (1987). *Modern NMR Spectroscopy: A Guide for Chemists*, Oxford University Press, Oxford, U.K.
4. Chandrakumar, N. and Subramanian, S. (1987). *Modern Technique in High-resolution FT-NMR*, Springer, New York.
5. Brey, W.S. (Ed.) (1988). *Pulse Methods in 1D and 2D liquid–phase NMR*, Academic Press, San Diego.
6. Claridge Timothy D.W. (2009). High-resolution NMR technique in organic chemistry, Elsevier, Oxford.

Carbon-13 Nuclear Magnetic Resonance Spectroscopy

The low natural abundance of ^{13}C (1.10%) and lower gyromagnetic ratio than that of proton requires greater sensitivity for recording CMR in comparison to PMR of an organic compound. The availability of mini computers has made feasible the regular use of pulse methods in NMR spectroscopy. The high sensitivity of pulse method related with Fourier-Transform techniques of mathematical analysis have made feasible the regular acquisition of the NMR spectra of low abundant nuclei, such as ^{13}C, ^{15}N, etc. The pulse FT method also allows the routine measurement of an important parameter, the spin–lattice relaxation times of the nuclei which provide information of the dynamics of the molecule (Soon Ng). It is feasible to have greater resolution for the ^{13}C spectrum due to spreading of carbon resonances over a range of about 0 – 220 ppm in comparison to 0 – 20 ppm for the protons. The non-equivalent nuclei usually give separate absorptions, unlike protons which commonly contribute overlapping absorptions.

Usually, three items of information, namely, chemical shift, coupling constants and relative number of absorbing nuclei are obtained from the PMR spectrum, while the ^{13}C spectrum give routinely the chemical shift. It is normally not possible to determine the relative number of absorbing nuclei by measuring the peak area and also the quantitative coupling constants. The large ^{13}C–^{1}H coupling constants varying from 125–200 Hz for directly bonded protons consequence in multiplets which overlap to substantial extent giving a complex spectrum which is difficult to be analysed in the absence of decoupling. Thus, each absorption in ^{13}C spectra appears as a singlet due to spin decoupling caused by wide band —or noise decoupling technique resulting in increase of sensitivity but loss of spin-spin pattern. The coupling of directly bonded carbon and hydrogen is observed by off-resonance decoupling method but the separation of peaks produced using the method is not equal to ^{13}C–^{1}H coupling constant. Thus, the carbon atoms related with $-CH_3$. $-CH_2$, and $-CH$ groups can be identified due to appearance of absorption as quartets, triplets and doublets separately for the equivalent bonded hydrogen atoms.

The chemical shift values of carbon absorptions are expressed by taking tetramethyl silane (TMS) absorption as the reference value. The positive values of chemical shift increases towards lower field side similar to δ scale in PMR spectroscopy. Most of the absorptions appears upto 220 ppm with the occurrence of methyl absorptions and carbonyl absorptions at higher field side and lower field side respectively. It has been observed that the hybridisation of the carbon atom has a considerable influence on the chemical shift. The sp^3–hybridised C atoms absorbs at high field (0–60 ppm down field from TMS), sp^2–carbon at low field (80–200 ppm) and sp-carbon atom at intermediate values (Wehrli and Wirthlin, 1978; Levy et al., 1978). The carbon-13 spectroscopy is useful in the structure elucidation of organic compounds and also have number of other application (Soon Ng).

The detailed description about carbon-13 spectroscopy has been given in a number of books of which the names of two are given below.

1. Wehrli, F.W. and Wirthlin, T. (1978). Interpretation of C-13 NMR spectra, Heyden, London.
2. Levy, G.C., Lichter, R.L. and Nelson, G.L. (1980). Carbon-13 NMR, 2nd Edn, John Wiley, New York.

Mass Spectrometry

The molecules of a chemical compound are bombarded with a beam of electrons in a mass spectrometer under vacuum. The electron beam knock other electrons out of the molecule, which are thus converted to cation radicals (molecular ions) (Finar, 1975, reprinted 1996).

$$M + e \longrightarrow M^{+} + 2e$$

The positive ion, M^+ known as molecular ion peak and is produced when the energy of the electrons is equal to that of ionisation potential (normally 10–15 electron volts). The electron energy (50–70 eV) is able to form molecular ion with an excess of energy, which is sufficient to break it into a mixture of positively charged and neutral fragments, and can also undergo further fragmentation depending upon the remaining energy. The majority of the ions carry a unit positive, but some may carry a double (or greater) positive charge and a few may possess a negative charge by electron capture. All positive ions are accelerated in an electric field followed by a magnetic field. The ions having same mass/charge (m/e) ratio are collected into beams, and drop down on a collector plate. The ions are thus separated according to their mass/charge ratios and it is possible to determine the mass of the ions. As most of the ions carry a positive charge, m/e is equal to *m*. Very high resolution mass spectrometers are available to differentiate masses in the third decimal place. The mass spectrum of a compound is a plot of intensities of ion beam (y-axis) against mass/charge (m/e) and is distinctive of a pure compound. The reproducible mass spectra may be obtained by using the same conditions and instrument. The highest peak in mass spectrum indicates it to be a most abundant ion and is identified as the base peak (arbitrarily, 100%) and all other peaks are reported in comparison to this peak. The different compounds have different fragmentation pattern. Mass spectroscopy (MS) is frequently used in conjunction with GLC, and the combined operation provides at one go a qualitative and quantitative identification of the many structurally complex components that may be present together in a particular plant extract, e.g., essential oil (Harborne, 1998). In GC-MS spectra, the peaks are characterised by comparing their their mass spectra with mass library. A compound has a characteristic mass spectra and can be characterised by its comparison from the mass library.

Some important mass spectrometric ionisation techniques used for the determination of the molecular ions are described below.

(a) Electron Ionisation (EI)
(b) Chemical Ionisation (CI)
(c) Fast Atom Bombardment (FAB)

The details are available in a number of books.

The GC-MS technique is particularly useful for the identification of chemical compounds present in the essential oils, in a mixture of methyl esters of fatty acids and other complex mixtures containing volatile compounds by comparing their mass spectrum with standard mass library. The HPLC-MS is also very useful for analysing the thermally labile, high polarity or high molecular mass organic compounds.

4
CHAPTER

Chemicals from Plants

INTRODUCTION

Numerous plant species occur in nature and are essential for survival of humans and animals on the earth in the world. About 45,500 plant species have been identified in India (Anon, 2009; Bawa et al., 2009). These plant species are the sources of useful products to the people for their existence on the earth. The products consist of carbohydrates, medicines, fatty oils, essential oils, pesticides, tannins and dyes, gums, resins and oleo-resins, fibres and flosses, bamboos and canes, and a numbers of other products besides major products, viz., Wood and fuel wood. The wood of plants consists mainly of cellulose, hemicellulose and lignins. The different parts of the plants also contain a number of other chemical compounds known as secondary metabolites, and their presence is responsible for their medicinal, and other properties. Secondary compounds may be defined as those natural products, usually of plant origin, which do not function directly in the primary biochemical activities that support the growth, development and reproduction of the organism in which they occur. A number of compounds belonging to different classes from alkaloids to terpenoids and acetogenins to phenolics and others have been isolated from different plant species using various separation technique (CC, TLC, PC, HPLC and others) and characterised on the basis of their spectral data (IR, UV, NMR, MS), preparation of derivatives and degradative studies. Isolation and structure elucidation of compounds from different parts of the plants can be accomplished as given below.

ISOLATION AND CHARACTERISATION OF CHEMICAL CONSTITUENTS OF PLANTS

Plant Selection

All the plant species existing on the earth are useful and sometimes a single plant species may be of multipurpose use. Therefore, it is important to select a plant species according to the subject of interest to the researcher. The plant species yielding useful products of high utilities may be taken for further research work as per the requirement. It is always advisable to the students working for their Doctor of Philosophy Degree, to explore the chemical compounds of unexplored or comparatively lesser chemically investigated plant species. The probability of discovering novel compounds is more and the research results can be published in the standard natural products chemistry journals. The already known compounds isolated from the plant species under chemical examination are of academic interest only. Therefore, the selection of plant species is very important for the researchers in the field of natural products chemistry. The synthetic confirmation for the structure of the characterised compounds may also be provided.

The selection of a plant species can be made by knowing its medicinal and other properties from the literature, viz., The Wealth of India and other related books. The review of chemical literature on a particular plant species can be done mainly from science finder, and chemical abstracts which will provide information about the number and classes of compounds isolated and characterised earlier from the particular plant species and genus. This information is essential before taking the research work on a plant species. Therefore, it is always advisable to select a hitherto unexplored plant species for the doctoral work. The established researchers may take up the research work on plants possessing medicinal and other activities for the development of value added herbal products which are in demand at present.

Collection of the Plant Material

The plant parts which possess varying medicinal and other activities, and are used for various purposes, are to be collected for their chemical examination in order to characterise the compounds responsible for these activities either alone or in combination with other compounds (synergetic effect). The statistically designed method may be used for the collection of the plant material to be taken for the chemical examination in order to obtain the average yield of the compounds present in them. It is usually advisable to collect the plant material from fully grown plants and approximately of the same age. There may be seasonal and geographical variation in the chemical constituents of a plant species.

Processing of Plant Material

The plant material after collection may be chopped into small pieces and air dried in shade followed by its pulverisation to a normally desired mesh size (5–20) for its efficient extraction with a solvent to get the maximum yield of extract. There may be problems with finer mesh size of the plant material as it will retain the considerable amount of the solvent used for extraction. There will be bumping of the solvent in the extraction flask in case of direct extraction of the plant material.

Extraction of Plant Material

The plant parts of different plant species contain the compounds belonging to different classes of compounds, namely, alkaloids, carbohydrates, coumarins, essential oils, fatty oils, flavonoids, phenolic compounds, tannins, terpenoids, and other classes depending on the family of plant species taken for examination. These may be non-polar, polar and highly polar compounds. Their isolation from the plant parts is usually accomplished by extracting with non-polar, polar or highly polar solvents. Hexane, petroleum ether, benzene normally come under the first category, while chloroform, acetone, ethyl acetate, are covered under the second category. The pure methanol or ethanol and containing 5–10 per cent or higher proportions of water, and pure water belong to category of highly polar solvents. The supercritical carbon dioxide is the latest non-polar solvent used for the extraction of essential oil and other non-polar compounds and its blending with (5–10% methanol) may be used for the extraction of relatively polar compounds. The solvents of increasing polarity normally used for the extraction of compounds from plant materials are – petroleum ether, acetone, methanol, or ethanol, varying percentage of water in methanol or ethanol. There are other solvents of varying polarity which fall in between these solvents. The sequential extraction of the plant material is usually carried out to obtain the extracts of varying polarity.

The extraction with petroleum ether, is carried out to isolate waxes, hydrocarbons, chlorophylls, essential oils, fatty oils, and other non-polar compounds. The extract yield is normally on the lower side except for the fatty oils. The acetone is usually used for extracting the polar compounds, viz., phenolic acids, flavonoids, triterpenoids, coumarins, alkaloids, and other plant constituents. The extract yield is normally more than that obtained using petroleum ether. The highly polar compounds like saponins, flavonoids glycosides, phenolic glycosides, tannins, alkaloids, polysaccharides, lignins, and other polar plants constituents may be extracted with methanol, ethanol, either alone or containing varying percentage of water. The extract yield is generally on the higher side in comparison to the petroleum ether and acetone extracts.

Thus the sequential extraction of a plant material is usually carried out as given below.

(i) The plant material is extracted with petroleum ether.

(ii) The left out plant material is again extracted with acetone.

(iii) The plant material left after extraction with acetone is again extracted with ethanol or methanol. It may also be further extracted with methanol or ethanol-water mixture or with pure water as per the requirement.

(iv) The solvent may be recovered from the extracts on a water bath or on a rotary vaccum flash evaporator.

Examination of Extracts

The extracts obtained consists of complex mixture of secondary metabolites present in the plant material and usually examined for their number by Thin Layer Chromatography (TLC) in the solvents of appropriate polarity. The plate is normally sprayed with (5% sulphuric acid) followed by heating in an oven at about 80-90°C to detect the approximate number of compounds-present in the extract. The probable nature of compounds may be found by spraying the TLC plates with different spray reagents as given in Table 4.1. A number of other spray reagents are also used for the spots development on the TLC plates.

TABLE 4.1

S. No.	Spray Reagent	Colour of Spots	Probable Nature of Organic Compounds
1.	5% alcoholic ferric chloride	Blue, green and brown	Presence of phenolic hydroxyls in the compound
2.	5% sulphuric acid	Pink, Red and Violet colour on heating. Yellow colour on heating.	Triterpenoids, steroids, chalcone Flavone, flavonol
3.	Dragendroff reagent	Orange	Alkaloids

The extracts consist of a number of compounds belonging to different classes of organic compounds. In case of a complex extract it is advisable to simplify it by maceration with suitable organic solvents. For example, petroleum ether can be used for the maceration of chloroform extract, while ether and ethyl acetate are suitable for the maceration of methanol, ethanol and water extracts. The sequential maceration with ether, ethyl acetate, and methanol may be carried out for solid water extract. The methanol, ethanol or water extracts containing polar compounds may be fractionated by suspending the solid extract in water in appropriate material (extract) to water ratio and extracting sequentially with ether, ethyl acetate and *n*-butanol using a separating funnel. The *n*-butanol exract contains mainly the very polar compounds namely saponins, flavonoid glycosides, phenolic compounds glycosides and other very polar compounds. The free sugars present in the extracts become soluble in water and can be separated there by simplifying the extracts. Isolation of pure compounds from the extracts can be carried out by their column chromatography over silica gel followed by elution of the column with solvents of increasing polarity starting from petroleum ether, mixture of petroleum ether with chloroform, pure chloroform, and varying proportion of methanol in chloroform. The other adsorbents like alumina, sephadex, and others may also be used for column chromatography according to the nature of compounds to be isolated. The choice of eluting solvent can be made according to the nature of adsorbent used and compounds to be isolated.

GENERAL PROCEDURE FOR COLUMN CHROMATOGRAPHY OF EXTRACTS

(i) Dissolve the extract in minimum quantity of a suitable solvent in which it is readily soluble.

(ii) Adsorb the extract on the optimum quantity of adsorbent, for example, silica gel.

(iii) Remove the solvent under vacuum by using a rotary flash evaporator at room temperature.

(iv) Make the slurry of adsorbent (silica gel) in a solvent in which the column is to be built and can be known by studying the TLC behaviour of the extract. The column of appropriate length, depending on the quantity of extract, is required to be built for the separation of the compounds. The extract to adsorbent ratio should be appropriate and precautions to be taken so that the former is not overloaded on the later.

(v) The extract adsorbed on silica gel is charged on top of the column.

(vi) The retention volume of the column should be determined for changing the polarity of eluent. It can be determined by subtracting the volume of the solvent obtained after charging the column from the initial volume taken for building the column.

(vii) The fractions of a fixed volume, as per size of the column and the amount of extract adsorbed on silica gel for fractionation, may be collected to isolate the compounds.

(viii) The solvent may be removed on a water bath or on a rotary vaccum flash evaporator under reduced pressure to isolate the eluted compounds as a mixture or in pure form.

(ix) Thin Layer Chromatography (TLC) of the fractions obtained can be performed in a pure solvent or a mixture of solvents of appropriate polarity depending on their movement, to know the number of compounds present in them. The similar fractions may be mixed together to process them further for the isolation of pure compounds.

(x) The elution of the column is to be continued till the removal of all the compounds of the extract duly monitored by TLC.

(xi) The pure compounds may be isolated either by crystallisation or by repeated column chromatography.

(xii) It is advisable to complete the column chromatography in the shortest possible time. The extended time may lead to retention of some compounds or some minor changes in a few of the fragile compounds due to their prolonged presence on the adsorbent (silica gel) of the column.

Column Chromatography of Petroleum Ether Extract

The petroleum ether extract is usually adsorbed on silica gel or any other adsorbent depending on the nature of compounds to be separated. The column of an appropriate length is built using silica gel and petroleum ether as solvent. The adsorbed material is charged on the top of the column and elution of the column with petroleum ether followed by fraction collection of a particular volume which can be increased or decreased as per the requirement. The polarity of eluting solvent (petroleum ether) may be changed by using varying amount of chloroform in petroleum ether, chloroform, and varying amount of methanol in chloroform as per the need. The fractions obtained are examined on TLC followed by mixing of similar fractions to make a major fraction. The factions collected may be processed for the isolation of pure compounds either by crystallisation or by repeated column chromatography depending on the complexity of the fractions. The non-polar compounds, viz., mono, sesqui or diterpenoids, sterols, simple triterpenes, long chain alcohols, fatty acids, methylated flavonoids, etc., are generally isolated from the petroleum ether extract.

Column Chromatography of Acetone Extract

The column chromatography of acetone extract may be done by following the similar procedure as given above for the petroleum ether extract. The adsorbent slurry may be prepared in the solvent of appropriate polarity as indicated by the study of TLC behaviour of the extract to be column chromatographed. For example, if the extract exhibits movement of compounds on TLC in 5 per cent methanol-chloroform, the silica gel slurry may be prepared in chloroform for building the column. The polarity of the solvent may be changed by increasing the percentage of methanol. It is not advisable to start the column with petroleum ether as it will take long time for its completion.

Column Chromatography of Methanol Extract

The methanol extract of a plant material normally contains a complex mixture of compounds as compared to petroleum ether and acetone extracts. It can be simplified to some extent either by maceration (with ether, ethyl acetate) or by suspending it in water and fractionating with ether, ethyl acetate and *n*-butanol sequentially by using a separating funnel. Thus, the methanol extract as such or fractionated methanol extract fractions (ether, ethyl acetate and *n*-butanol soluble fractions) may be put to column chromatography by following the similar procedure as described for the acetone extract. The solvent for making the silica gel slurry for building the column may be chosen by studying its TLC behaviour in order to complete the column in shortest possible time. For example, if the movement of compounds of an extract is observed in ethyl acetate-methanol–water (6:4:3), then the column may be built directly in chloroform or 2% methanol-chloroform mixture. The column may be eluted with increasing percentage of methanol in chloroform as per the movements of compounds in the column, monitored by TLC in appropriate solvent system and using appropriate spray reagents. Normally, most of the compounds present in the methanol extract are eluted up to 30 per cent of methanol in chloroform. Elution with higher percentage of methanol in chloroform may dissolve silica gel and get it eluted in the fractions of the column. Therefore, the researcher has to be careful while eluting the column with higher percentage (30–40%) of methanol-chloroform mixture. Usually, the column is not eluted with pure methanol.

The simultaneous examination of the column fractions should be done in order to find the number of compounds present in them and for further work up by mixing the similar fractions. Depending upon the complexity of fractions, the pure compounds can either be isolated by crystallisation or by repeated column chromatography mostly over silica gel or other adsorbents as per the nature of compounds to be isolated. The fractions consisting of small amount of two compounds can be separated by preparative TLC, in which the mixture is applied as a continuous band at the bottom of the plate which could be developed in a suitable solvent in which the two bands separate with ease. The single band can be observed either by exposing the plate to UV light or to iodine vapours and marked bands can be removed separately from the plates. The pure compound can be isolated by charging them in a small column and eluting with a solvent which can quickly remove it from the column. The compounds can also be isolated by preparative HPLC depending upon the availability of the instrument.

When a compound shows a single spot on TLC in different solvent systems and possesses a sharp melting point, it is usually taken to be pure. The purity can also be tested by HPLC and if volatile by GLC. The purity of a compound is essential for its structure determination. Impure compounds lead to erratic results.

STRUCTURE ELUCIDATION OF A PURE COMPOUND

There has been a steady increase in the isolation and structure elucidation of compounds from the plants due to the continuous developments in the field of chromatography and spectroscopy. During the first half of twentieth century, measuring of some physical properties, chemical tests, derivatisation, chemical degradation of compounds were mainly used for the structure elucidation and comparatively a lesser number of compounds from the plants were discovered during that period. But due to development and application of chromatographic techniques (CC, TLC, PC, GLC and HPLC) for the separation of compounds and spectroscopic techniques (IR, UV, NMR and MS) for their structure elucidation, a large number of compounds were discovered during the second half of twentieth century. Further developments in field of separation techniques and spectroscopy like GC-MS, HPLC-MS, 2DNMR and other sophisticated techniques have made it possible to separate and characterise a large number of minor organic compounds present in the plants. The process of structure elucidation is fast and precise. Therefore, numerous compounds with unique structures are being discovered every year from the plants which strengthens the development in the field of organic chemistry. The general procedures which may be used for the structure elucidation of a pure compound are given below.

Determination of Melting Point

The melting point of a pure compound is determined in the beginning of its structure elucidation. The low melting point is usually observed for long chain compounds, fatty acids, non-polar and other compounds. The polar compounds usually possess higher melting point. There are compounds which do not melt but decompose on heating. These are some of my general observations, and cannot be correlated with the structures of the compounds.

Determination of Class of Compound

The nature of a compound may be determined by performing some chemical colour tests, viz., Liebermann-Burchard test for triterpenoids, Salkowski test for triterpenoids/steroids (Finar, 1975, reprinted 1976), Shinoda test and Pew test for flavonoids (flavones/flavonols) (Shinoda, 1928; Pew, 1948), alcoholic ferric chloride spray test for the presence of phenolic groups, Molisch test for sugars (glycosides), Maeyer's test and Dragendorff test for alkaloids. The other tests have also been described in the literature for different classes of compounds.

Infrared Spectra

The infrared spectrum of a compound is recorded to detect the presence of important functional groups, viz., –OH, –COOH, $>C=O$, $-NH_2$, $-O.COCH_3$, $-COOCH_3$, or patterns of substitution on double bonds or aromatic rings from the spectra. The high frequency end (4000–1500 cm^{-1}) of the IR spectra is useful for the identification of functional groups due to relatively low overlapping of the peaks in this area. The absorption at ~3300 cm^{-1} and 1700 cm^{-1} generally indicate the presence of –OH and $>C=O$ peaks, respectively. The IR spectrum with frequency below 1500 cm^{-1} contains numerous peaks due to overlapping, stretching, and bending absorptions and known as finger print region (1400–650 cm^{-1}). This region is specific of a compound as it contains a great deal of information about the pattern of substitution in the structure of a compound instead the types of functional groups. The finger print region is thus characteristic of a compound and the superimposable IR spectrum of an unknown compound under investigation (say suspecting to be ursolic acid Fig. 4.1) can be recorded with an authentic sample of ursolic acid to confirm the identity of the compound under examination.

Ultraviolet Spectra

The UV spectra of organic compounds helps in their structure elucidation. The UV spectrum of a flavonoid in methanol or ethanol is very informative. It indicates the compound to be a flavones, flavonol, isoflavones, chalcone or flavonone. The anthocyanins have a characteristic spectrum in the visible region of the spectrum. The recording of UV spectrum with shifts reagents like aluminium chloride, aluminium chloride/hydrochloric acid, sodium acetate, sodium acetate/boric acid, sodium methoxide, indicate the presence of hydroxyl groups at different positions in the flavonoids. The details about characteristic shifts have been described in a number of books (Mabry et al., 1975). The brief description about ultraviolet visible spectroscopy have been given in chapter 3.

Preparation of Derivatives

The recording of IR spectrum indicates the presence of hydroxyl (–OH), carboxyl (–COOH), and other functional groups in the compounds. The number of hydroxyl groups (alcoholic or phenolic) present in the compound may be determined by preparing the acetate of the compound with acetic anhydride/pyridine, phenolic hydroxyls also by preparing methyl ether with dimethyl sulphate/potassium carbonate, carboxyl group by preparing the methyl ester of the compound with diazomethane followed by recording of its NMR spectrum. The presence of a number of alcoholic acetoxyls, phenolic acetoxyls, and methyl ether groups will indicate the presence of same number of alcoholic hydroxyls and phenolic hydroxyls groups in the compound, respectively. The tertiary hydroxyl group does not undergo acetylation. Similarly, a number of other derivatives may be prepared depending on the availability of the compound and presence of functional groups in it. It is now possible to determine the structure of small amount of the compound with advance spectroscopic techniques without preparing the derivatives.

Nuclear Magnetic Resonance Spectra

The nuclear magnetic resonance (NMR) spectra of organic compounds are very informative for their structure elucidation. The main advantage of NMR spectra, for the structure elucidation of chemical compounds from a plant species is that it yields a complete picture of molecular structure and behaviour in solution. Besides establishing the primary structure of the organic compound, NMR also provides information on the conformation and molecular dynamics (molecular characteristics) of the molecule in the solution state. It is now possible to record the NMR spectra (^{1}H,^{13}C) with small amount of a compound due to availability of high field NMR instruments. NMR is a non-destructive technique for the organic compounds (Agrawal, 1994). It provides the precise information about the chemical environment of hydrogen and carbon atoms.

The presence and number of methoxyls (–OCH3), alcoholic and phenolic acetoxyls (–O.$COCH_3$), methyl, methylene, methine groups, aromatic and other protons in a compound can be determined by its ^{1}HNMR. The chemical environment of carbon atoms in an organic compound can be determined by recording its ^{13}CNMR spectrum.

The introduction of Fourier Transform (FT) technique in nuclear magnetic resonance (NMR) spectroscopy by Ernst and Anderson in 1966 announced a new era in this branch of spectroscopy (Ernst and Anderson, 1966). The NMR spectrum of a compound can also be recorded as one- and two-dimensional NMR. The structure elucidation of a natural product require a variety of NMR experiments to be performed for the complete assignment of proton and carbon chemical shifts. The information which can be obtained from different one- and two-dimensional NMR experiments have been described (King and William, 1990; Agrawal, 1994).

(1) *One-dimensional NMR Techniques (1D)*

(a) ^{1}HNMR— It gives information about chemical shift, coupling constants, number of protons and vicinities in space (NOE difference).

(b) ^{13}CNMR—The information about chemical shift and number of attached hydrogen atoms (APT and DEPT) is provided.

(2) *Two-dimensional NMR techniques (2D)*

(a) *Homonuclear Spectroscopy:*

(i) ^{1}H-^{1}H COSY—Information about ^{1}H, ^{1}H correlation, identification of coupling partner, chemical shift and coupling constants (PS/DQF) is provided.

(ii) TQF-COSY—Information about ^{1}H, ^{1}H correlation, for three and more mutually coupled spin system, chemical shift is provided.

(iii) GEM-COSY—Provide information about ^{1}H, ^{1}H correlation for germinal methylene pairs.

(iv) TOSCY/RELAY/HOHAHA—Provide information about ^{1}H, ^{1}H correlation to intra-residual protons pertaining to same J-network.

(v) NOESY/ROESY—Provide information about ^{1}H, ^{1}H correlation through space.

(vi) 2D J-Spectroscopy—Provide information about ^{1}H chemical shift and ^{1}H–^{1}H coupling constants.

(vii) INADEQUATE—Provide information about ^{13}C–^{13}C connectivity and detection of directly bonded carbon atoms.

(b) *Heteronuclear Spectroscopy:*

(i) ^{1}H-^{13}C COSY (HETCOR/HMQC)—Information about one-bonded correlation between directly bonded ^{1}H and ^{13}C chemical shifts is provided.

(ii) Long–range ^{1}H-^{13}C COSY (LR-HETCOR) (HMBC)—It provides information about two and three bond ^{1}H-^{13}C correlation and detection of long range coupling partner containing quaternary carbon atoms.

(iii) 2D J spectroscopy—Information about ^{13}C and ^{1}H chemical shift, $^{1}J_{CH}$ is provided.

(iv) HMQC-COSY—Provides information about one bonded ^{1}H-^{13}C correlation, $^{1}J_{CH}$ and detection of vicinal proton resonance.

(v) HMQC-TOCSY—Provides information about one bonded ^{1}H-^{13}C correlation and detection of vicinal and other proton resonance.

(vi) HMQC-NOESY—Provides information about one bonded ^{1}H-^{13}C correlation and detection of protons which are in close nearness.

It is possible to predict the structure of a compound on the basis of its noise free NMR spectra. The basic information about NMR spectroscopy have been described in chapter III. The details about the various aspects of NMR spectroscopy have been described in a number of books.

Mass Spectra

The molecules of a chemical compound are bombarded with a beam of electrons in a mass spectrometer under vacuum. The electron beam knock the other electrons out of the molecules, which are thus converted to cation radicals (molecular ions). The molecular ions incline to break down immediately into smaller fragments, and these in turn tends to break down into even smaller pieces. The fragments are separated from one another by exposing them to powerful magnetic field in the mass spectrometer leading to their deflection from their normal line of flight into a detector. The uncharged fragments are not affected by magnetic field. The deflection of particles is proportional to the charge on each fragment and inversely proportional to its mass. Thus, the separation of particles by the mass spectrometer into separate peaks take place on the basis of mass-to-charge ratio. The mass spectrum contains a number of peaks arranged in order of mass because most of the particles carry a single positive charge. Therefore a mass spectrum is different from spectra like IR, UV and NMR. The most common use of a mass spectrum is to determine molecular weight of the compound or molecule.

The molecular ions are produced by the loss of electrons from the parent molecules. The mass of molecular ion measured by mass spectrometer is the same as the molecular weight of the molecule because of essentially weightless electrons. The characteristic fragments are obtained from a compound, and thus their interpretation provide valuable information about its structure. The mass spectra are useful for determination of the structure of the compound isolated or occurring in very small amounts, viz., the sex attractants of insects which occur in trace amounts. Each compound has a characteristic mass spectrum with specific fragmentation pattern and its structure can be elucidated by using its mass spectrum and other spectral data (IR, UV and NMR).

The GC-MS technique is particularly useful for the identification of chemical compounds present in the essential oils, in a mixture of methyl esters of fatty acids and other complex mixtures containing volatile compounds by comparing their mass spectrum with standard mass library. The high performance liquid chromatography-mass spectrometry (HPLC-MS) is also useful for the separation and identification of organic compounds present in a mixture. HPLC-MS is helpful in analysing the thermally labile, high polarity or high molecular mass organic compounds.

The details about mass spectrometry and its application for the structure elucidation of organic compounds have been described in a number of books by different authors. The book entitled "Spectrometric identification of organic compounds" by Silverstein et al. is one of fundamental books to know about the basic information about infrared spectroscopy, ultraviolet and visible spectroscopy, nuclear magnetic resonance spectroscopy, and mass spectrometry.

Identification of Sugars

The presence of sugars in a compound is indicated by its positive Molisch test. The identification of sugars in a compound can be done by its hydrolysis with Kiliani mixture or hydrochloric acid (5%). The aglycone can be separated from the reaction mixture by extracting with an organic solvent like ether, ethyl acetate or chloroform depending on its solubility. The acidic portion of the reaction mixture is neutralised with barium carbonate, filtered and evaporated in vacuum to obtain the sugar portion of the glycoside. Its paper chromatography in the solvent system like butanol-pyridine-water (6:4:3) or other solvent systems along with authentic sugar samples can be done to characterise the different sugars present in the compound. The solvent may be removed from the paper by drying in the air, sprayed with aniline hydrogen phthalate reagent, and heated in the oven for spot development. The sample spot having an identical Rf value with an authentic sugar sample can be identified to be that sugar moiety. The other reagents can also be used for the spots development.

The permethylation of a glycoside using Hakomori's method and hydrolysis of the methylated glycoside followed by paper chromatography of the methylated sugar portion was found to be helpful in the structure elucidation of the glycoside (Dayal and Parthasarathy, 1977; Dayal, 1976). It is now possible to characterize the sugar moieties by ^{13}CNMR of the glycosides (Mendiratta et al., 2007)

STRUCTURE ELUCIDATION OF TRITERPENOIDS

The triterpenoids/steroids occur in a number of plants species. These are present as sapogenins (without sugar moiety) and saponins (with a sugar moiety or moieties). The former may be extracted with petroleum ether, chloroform, acetone depending on the presence of no functional groups or functional groups like hydroxyl, carboxyl, aldehyde, ketone or other groups in them. For example, the triterpenes without functional groups may be extracted with petroleum ether, while those with hydroxyl and carboxyl with chloroform or acetone. The saponins are usually extracted with methanol or ethanol and those containing more than three sugar moieties with higher polarity solvents like 5–10 per cent of water in methanol. The methanol, ethanol, or methanol-water mixture extracts may be fractionated into ether, ethyl acetate, and *n*-butanol soluble portions after suspending them in water in a separating funnel. The ether soluble portion may contain sapogenins or other non-glycosidic compounds. The ethyl acetate soluble portion usually contains the sapogenins or sometime saponins containing a sugar moiety. The *n*-butanol soluble portion consists of a mixture of saponins. The presence of saponins can be indicated by shaking a small amount of extract with water in a test tube which will lead to excessive frothing in the test tube and which may sustain for about 12 hours or more.

The pure compounds may be isolated by column chromatography usually over silica gel (100–200 mesh) and the eluant may be selected by studying the TLC behaviour of the extract. For example, if the spots movement on TLC is observed in methanol-chloroform 5 per cent, then the column may be built in chloroform and the elution of the column may be started with the same solvent. The volume of the fractions to be collected is decided after taking into consideration the column size and the amount of extract taken for the column chromatography. The fraction volume may generally vary from 100–300 ml except in small columns. The eluted compounds may be collected after the removal of solvent on a water bath or rotary vacuum flash evaporator. The fractions may be examined on TLC in the solvent system of appropriate polarity. For example, the compounds eluted with 5 per cent of methanol–chloroform from the column may move in about 6–10% of methanol-chloroform on TLC, probably due to presence of binder in the TLC brand silica gel. The spots can be developed by spraying the TLC plates with about 5 percent sulphuric acid. The triterpenes spots usually appear as red, pink and violet on heating the plates at about 90°C in oven. It has been observed that beta-sitosterol spot appears as red on heating which finally turns green on further heating or by keeping the plate in air. The ursolic acid spot appears as red to pink. The fractions showing similar spots on TLC may be mixed together and then either crystallised or recolumn chromatograhed to isolate pure compounds. The pure compound should show a single spot on TLC in at least two solvent systems and possess sharp melting point. Sometimes two similar compounds having the variation in double bond position in the molecule may show a single spot on TLC. Therefore, the purity of a triterpene may be ascertained by its TLC on silver nitrate impregnated plate. The resolution of this type of compounds is feasible by using these plates. The two compounds may be separated by column chromatography over silver nitrate impregnated silica gel. The column may be covered with the black carbon paper to

(iv) Preparation of Derivatives: Having known the presence of phenolic hydroxyl groups in the compound, it advisable to prepare its derivatives, viz., acetate or methyl ether derivatives. The derivatives are readily soluble in deuterated solvents for recording their ^{1}HNMR spectrum for determining the number of hydroxyls and other protons in different chemical surroundings in the compound. The other derivatives of the compounds may also be prepared.

(v) NMR Spectrum: The ^{1}HNMR spectrum of a flavonoid derivative is comparatively more clear (noise-free) and informative than that of a parent compound. For example, the ^{1}HNMR of acetate of a compound gives clear indication about the presence of number of phenolic and alcoholic hydroxyl groups, methoxyl and methyl groups in the parent compound. The presence of ortho, meta- or para-aromatic protons can also be derived from the spectrum. The chemical surroundings of other hydrogen atoms present in the compound may also be predicted. The ^{13}CNMR spectrum of the compound gives information about the number of carbon atoms and their chemical surroundings, viz., attachment with hydrogen, oxygen in the compound which is very helpful in its structure elucidation. The 2D NMR of the compounds is also recorded which is very helpful in the structure elucidation.

(vi) Mass Spectrum (MS): The mass spectrum of the compound determines its molecular mass. The molecular ion peak (M^+) gives the molecular weight of the compound. The flavonoids belonging to different classes have characteristic fragmentation pattern which are helpful in determining their structures (Mabry et al., 1975). It is thus possible to elucidate the structure of flavonoid compounds by the collective use of IR, UV, NMR (^{1}HNMR, ^{13}CNMR, 2DNMR) and MS spectral data.

For example, the structure elucidation of artoflavanone, isolated from *Artocarpus heterophyllus* roots (Dayal and Seshadri, 1974), apigenin-5-*O*-β-D-glucopyranosyl-(2→1)-α-L-rhamnopyranoside and apigenin-5-*O*-β-D-(6″-acetylglucopyranosyl-(2→1) -α-L-rhamnopyranoside isolated from the needles of *Cephalotaxus harringtonia* var. *harringtonia* (Mendiratta Nee Chugh et al., 2007), on the basis of their spectral data is being given below.

(1) *Artoflavanone:* It was crystallised from petroleum ether as colourless needles, m.p.141-142°C; $C_{24}H_{28}O_7$;$[\alpha]_D$ +52°[c, 0.50 in $CHCl_3$]. The compound gave green colour on spraying the TLC plate with alcoholic ferric chloride. It also gave the characteristic test for flavanones with borohydride reagent (Krishnamurty and Seshadri, 1965). Its structure was elucidated as 6-C-prenyl-5-hydroxy-7, 3′, 4′, 5′-tetramethoxy flavanone (Artoflavanone) on the basis of spectral data (IR, UV, ^{1}HNMR and Mass) [Dayal and Seshadri, 1974].

The mass fragmentation of artoflavanone is shown in Fig. 4.2.

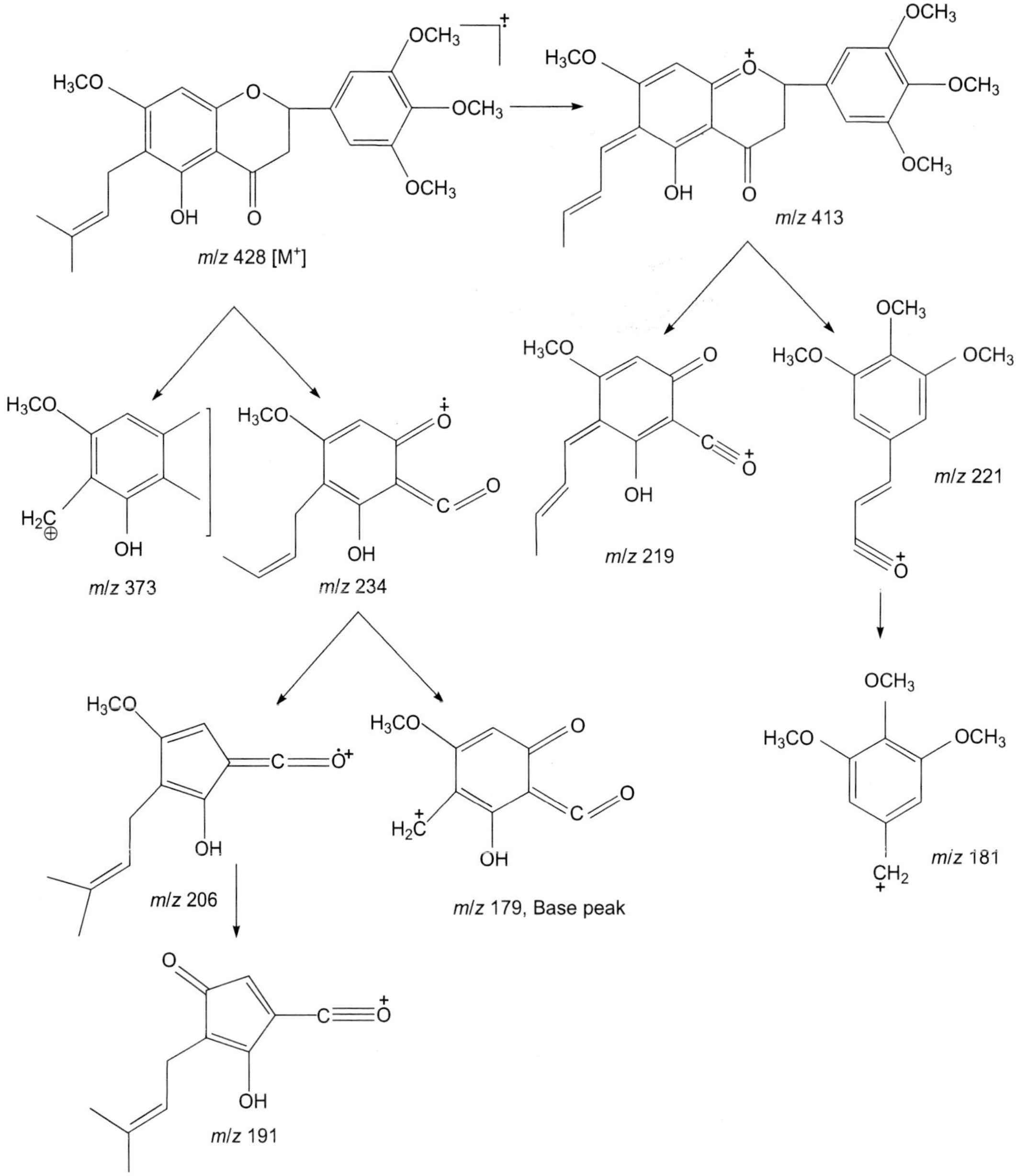

Fig. 4.2 Mass fragmentation pattern of artoflavanone

IR spectrum (ν, KBr): 3400(–OH), –1610 ($>$C=O), 1450, 1370, 1360, 1290, 1220, 1200, 1120, 1060, 950 and 810 cm^{-1}.

UV(λ_{max}, Methanol)

CH_3OH	290, 340 nm; (logε4.35, 4.00)
$CH_3OH+AlCl_3$	312, 397; nm
$CH_3OH+NaOAc$	290, 340; nm
$CH_3OH+NaOAc+H_3BO_3$	290 340; nm
$CH_3OH+NaOMe$	290, 365 nm.

^{1}HNMR spectrum ($CDCl_3$)

Position of signal (δ)	Number of protons	Multiplicity	Assignment
1.58; 1.65	6	2 Singlets	$=C(CH_3)_2$
2.35	1	Multiplet	C_3–H
2.60	1	Multiplet	C_3–H
3.20	2	Doublet, J = 7 Hz	$–CH_2–Ar$
3.98	12	Singlet	$–(OCH_3)_4$
5.15	1	Multiplet	C_2–H
5.85	1	Multiplet	Vinylic proton
6.10	1	Singlet	C_8–H
6.25	2	Singlet	C_2'–H and C_6'–H
12.28	1	Singlet	Chelated –OH

Mass spectrum (m/z) (% abundance):

428 (M^+, 78), 413(8), 373(24), 234(12), 221(15), 219(55), 206(33), 194(43), 191(63), 181(86), 179 (base peak, 100), 151(34), and 121(66).

The mass fragmentation pattern of artoflavanone is as given in Fig. 4.2.

(2) Apigenin-5-O-β-D-glucopyranosyl-(2→1)-α-L-rhamnopyranoside (Compound-1): The methanol extract was fractionated into benzene and ethyl acetate soluble portions and ethyl acetate insoluble portion. The ethyl acetate insoluble portion on column chromatography over silica gel yield two compounds 1 and 2. compound 1 was characterized as apigenin (IR, UV, ^{1}HNMR, ^{13}CNMR). Compound 2 was crystallized with methanol-chloroform as light yellow needless m.p. 260–265°C; $[\alpha]_D^{25}$: –70° (MeOH, c 0.1) which was characterized as apigenin-5-O-β-D-glucopyranosyl-(2→1)-α-L-rhamnopyranoside on the basis of spectral data as given below (Mendiratta (nee) Chugh et al., 2007; Mendiratta, 2003).

IR ν_{max}^{Nujol}: 3551, 1628, 1553, 1512, 1346, 1254, 1181, 833 and 723 cm^{-1}.

UV(λ_{max}^{MeOH} nm): 328.0, 262.5; + $AlCl_3$ 331.5, 262.0; +NaOAc 361.5, 269.0; +NaOAc+H_3BO_3 328.5, 263.0; +NaOMe 378.0, 269.0.

^{1}H NMR (500 MHz, DMSO-d_6): δ 7.82 (2 H, d, *J* = 8.7 Hz, H-2′, 6′), 6.91 (2 H, d, *J* = 8.7 Hz, H-3′, 5′), 6.57 (1 H, d, *J* = 2.5 Hz, H-8), 6.52 (1 H, d, *J* = 2.5 Hz, H-6), 6.45 (1 H, s, H-3), 5.29 (1 H, d, *J* = 6.4 Hz, H-1″), 5.19 (1 H, d, *J* = 1.5 Hz, H-1‴), 3.42-3.72 (10 H, m, sugar protons), 1.04 (3 H, d, *J* = 6.1 Hz, Rha-CH_3).

^{13}C NMR (125 MHz, C_5D_5N): δ 175.4 (C-4). 162.1 (C-7), 160.4 (C-4′), 160.1 (C-2), 158.8 (C-9), 157.5 (C-5), 127.7 (C-2′, 6′), 121.5 (C-1′), 115.9 (C-3′, 5′), 107.5 (C-10), 105.9 (C-3), 99.6 (C-1‴), 99.0 (C-6), 97.9 (C-1″), 96.2 (C-8), 77.0 (C-5″), 76.9 (C-2″, 3″), 72.3 (C-4‴), 72.0 (C-4″), 70.5 (C-2‴), 70.1 (C-3‴), 68.5 (C-5‴), 60.9 (C-6″), 17.8 (Rha-CH_3),

ESI-MS m/z: 579[M^+ + 1, 8], 409(7), 361(8), 288(9) and 271(8).

(3) Apigenin-5-O-β-D-(6″-acetylglucopyranosyl-(2→1)-α-L- rhamnopyranoside (Compound-2): The methanol extract was suspended in water and filtered. The aqueous methanolic extract suspension was partitioned between ethyl acetate and *n*-butanol to afford ethylacetate and *n*-butanol soluble fractions. The *n*-butanol soluble portion on repeated column chromatography over silica gel yielded compound 3 as a yellow solid m.p. 169-170°C $[\alpha]_D$: −100° (MeOH, c 0.1) and its structure was elucidated as apigenin-5-*O*-β-D-(6″-acetylglucopyranosyl-(2→1)-a-L-rhamnopyranoside on the basis of spectral data as described below (Mendiratta Nee Chugh et al., 2007).

IR ν_{max}^{Nujol}: 3349, 2724, 2363, 1721, 1636, 1601, 1563, 1291, 1250, 1183, 1146, 1088, 1063, 978, 837 and 723 cm^{-1}.

UV (λ_{max}^{MeOH}, nm) 327.0, 262.5; + $AlCl_3$ 328.5, 262.5; +NaOAc 368.0, 270.5; NaOAc+H_3BO_3 329.0, 262.0; NaOMe 380.0, 269.0.

^{1}H NMR (500 MHz, δ C_5D_5N): δ 7.79 (2 H, d, *J* = 9.0 Hz, H-2′, 6′), 7.18 (2 H, d, *J* = 9.0 Hz, H-3′, 5′), 6.71 (1 H, s, H-3), 7.13 (1 H, d, *J* = 2.5 Hz, H-6), 6.94 (1 H, d, *J* = 2.5 Hz, H-8), 6.37 (1 H, br s, H-1‴), 5.95 (1 H, d, *J* = 6.5 Hz, H-1″), 4.89 (1 H, m, H-5‴), 4.80 (1H,dd, *J* = 2.5, 11.8 Hz, H-$6''_b$), 4.68 (1 H, br s, H-2‴), 4.74 (1 H, m, H-2″), 4.65 (1 H, m, H-3‴), 4.61 (1 H, dd, *J* = 5.5, 11.8 Hz, H-$6''_a$), 4.41 (1 H, m, 5″), 4.39 (1 H, m, H-3″), 4.27 (1 H, m, H-4‴), 4.21 (1 H, m, H-4″), 1.88 (3 H, s, 6″—$OCOCH_3$), 1.64 (3 H, d, *J* = 6.0 Hz, Rha-CH_3).

^{13}C NMR (125 MHz, δ C_5D_5N): δ 177.2 (C-4), 170.7 (—$OCOCH_3$) 164.2 (C-7), 162.1 (C-4′), 161.5 (C-2), 160.0 (C-9), 158.6 (C-5), 128.5 (C-2′, 6′), 122.4 (C-1′), 116.7 (C-3′, 5′), 108.4 (C-10), 106.9 (C-3), 101.3 (C-1‴), 100.8 (C-6), 99.3 (C-1″), 97.5 (C-8), 78.9 (C-3″), 77.3 (C-2″), 75.0 (C-4″), 74.6 (C-4‴), 72.5 (C-2‴, 3‴), 71.5 (C-5″), 70.1 (C-5‴), 63.7 (C-6″), 20.5 (—$OCOCH_3$), 18.8 (Rha-CH_3).

FAB-MS m/z: 643(M^+ + Na, 2), 621(M^+ +1, 2), 460(6), 307(86), 289(53), 271(19), 219(9) and 165(24).

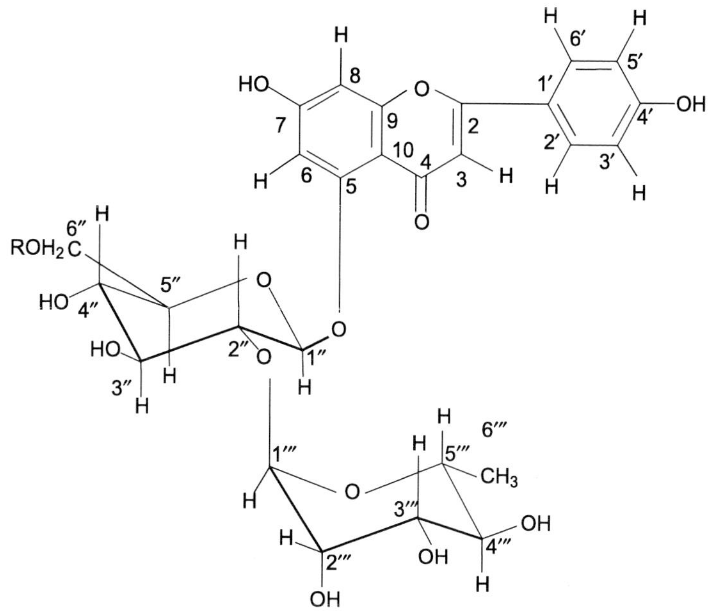

Compound 1 (R = H)
Compound 2 (R = Ac)

Fig. 4.3

The research team consisting of Dr. Rameshwar Dayal and other members carried out the research work on the isolation and characterisation of chemical constituents from a number of plant species which are given in Table 4.2.

Table 4.2 List of plant species examined for their chemical constituents

1.	*Artocarpus heterophyllus*
2.	*Asparagus racemosus*
3.	*Acacia catechu*
4.	*Achyranthes aspera*
5.	*Acorus calamus*
6.	*Adina cordifolia*
7.	*Cephalotaxus harringtonia*
8.	*Clematis roylei*
9.	*Cryptomeria japonica*
10.	*Dalbergia sericea*

Contd.

11.	*Dalberbia stipulacea*
12.	*Dicentra pauciflora*
13.	*Eucalyptus citriodora*
14.	*Eucalyptus globulus ssp globulus*
15.	*Eucalyptus* hybrid
16.	*Eucalyptus mealnophloia*
17.	*Gliricidia sepium*
18.	*Jacaranda mimosifolia*
19.	*Peucedanum grande*
20.	*Pinus roxburghii*
21.	*Rheum emodi*
22.	*Shorea robusta*
23.	*Syzinum cumini*
24.	*Tagetes minuta*
25.	*Tectona grandis*
26.	*Terminalia citrina*
27.	*Tetrameles nudiflora*
28.	*Vateria indica*
29.	*Vitex negundo*

ISOLATION OF ESSENTIAL OILS

The essential oils or volatile oils are present in the leaves, flowers, wood, and other parts of the plants. These oils consist mainly of mono- and sesquiterpenes and can be isolated from the plant parts by using a suitable method, viz., hydrodistillation, steam distillation, water-steam distillation, emuflerage, and by the latest supercritical fluid extraction method. The *Eucalyptus* oil isolated from the leaves of *Eucalyptus globulus,* Sandal wood oil from the *Santalum album* wood and *Tagetes* oil from the *Tagetes minuta* flowering twigs, are some of the examples. The essential oils are mainly used in perfumery, aromatherapy, and other applications. A detailed description about essential oils have been given in the Gunther's book volumes and other books on the essential oils (Gunther, 1948). The refractive index, optical rotation, specific gravity, acid value, ester value, are some of the physicochemical properties which are usually determined for the essential oils. Gas chromatography-Mass spectrometry (GC-MS) is an appropriate technique for the characterisation of chemical constituents of the essential oils. The research work on the essential oils of a number of plant species was carried out by a team consisting of author of this book and other members. The list of plants species from which the essential oils were isolated and characterised is being given in Table 4.3 and the details have been described in chapter 6.

Table 4.3 List of plant species examined for their essential oils

1.	*Achyranthes aspera* (Leaves)
2.	*Cephalotaxus harringtonia* var. *harringtonia* (Leaves and twigs)
3.	*Eucalyptus globulus* ssp. *bicostata* (Leaves)
4.	*Eucalyptus* hybrid (Leaves)
5.	*Eucalyptus tereticornis* x *Eucalyptus camaldulensis* (FRI-4) (Leaves), *Eucalyptus camaldulensis* x *Eucalyptus tereticornis* (FRI-5) (Leaves), *Eucalyptus tereticornis* (Leaves), *Eucalyptus camaldulensis* (Leaves), *Eucalyptus robusta* (Leaves), *Eucalyptus deglupta* (Leaves)
6.	*Pinus insularis* (Resin) and *Pinus merkusii* (Resin)
7.	*Shorea robusta* (Bast, Heartwood, Leaves, Resin)
8.	*Tagetes minuta* (Flowering twigs)
9.	*Vitex negundo* (Leaves and flowering twigs)

ISOLATION OF FATTY OILS

The fatty oils are isolated from the seeds either by using an expeller or by solvent extraction. The petroleum ether is usually used for the isolation of fatty oil from the powdered seeds using a Soxhlet apparatus. Removal of solvent yields the fatty oil which may be purified as per requirement. The fatty oil yield is higher using solvent extraction method. The large extractors are available for the solvent extraction of the seeds on industrial scale. The fatty oil yield is usually determined on moisture free basis of the seeds weight. The chemical composition of a fatty oil is determined by GLC of methyl esters of fatty acids obtained after its alkaline hydrolysis, followed by methylation of fatty acids with diazomethane or methanolic sulphuric acid and comparison with authentic samples of methyl esters. GC-MS of methyl esters may also used for the characterisation of fatty acids in the fatty oil. Palmitic, stearic, oleic, linoleic, linolenic acids are some of the major fatty acids which are present in the different fatty oils. The fatty oils have numerous uses. These are used as edible oils, for soap making and their methyl esters as biodesel. *Jatropha curcus* is being planted for the possible use of its oil as biodesel. Dayal et al., have carried out the research work on the chemical analysis of fatty oil isolated from the seeds of a number of plant species of forest origin in order to explore their possible use as biodesel. The complete list of plant species, the seeds of which were investigated for the fatty oil content, physicochemical properties of the oil and fatty acids present, is given in Table 4.4. Hydrolysis of fatty oils was carried out to isolate the mixture of fatty acids, which were esterified to obtain methyl esters. Fatty acids (%) in methyl esters prepared by esterification and transesterification were determined by GLC (Dayal and Thapliyal, 2006). The summarised details on the isolation and characterisation of fatty oils from some of the plant species, mentioned in Table 4.4, is being described in chapter 7.

TABLE 4.4 List of the plant species examined for fatty oils

S. No.	Name of species	S. No.	Name of species
1.	*Sapindus mukorossi*	15.	*Diospyros embroypteris*
2.	*Prinsepia utilis*	16.	*Quercus incana*
3.	*Heteropanax fragrans*	17.	*Pterospermum acerifolium*
4.	*Perilla frutescens*	18.	*Ardisia solanacea*
5.	*Bombax cieba*	19.	*Garuga pinnata*
6.	*Cordia myxa*	20.	*Lannea coromandelica*
7.	*Crataeva nurvala*	21.	*Diospyros montana*
8.	*Holoptelea integrifolia*	22.	*Diospyros melanoxylon*
9.	*Michelia champaca*	23.	*Sapium sebiferum*
10.	*Heynea trijuga*	24.	*Aleurites moluccana*
11.	*Putranjiva roxburghii*	25.	*Pongamia pinnata*
12.	*Mallotus philippensis*	26.	*Mesua ferrea*
13.	*Schleichera oleosa*	27.	*Prunus cerasoides*
14.	*Murraya koenigii*		

REFERENCES

Agrawal, P.K. (1994). Two-dimensional NMR spectroscopy in the structural elucidation of natural products, *Journal of Scientific and Industrial Research*, 53(5): 329–339.

Anon. (2009). *Training Module on Capacity Building of Communities Involved in Sustainable Forest Management*. Published by Forest Research Institute, Dehradun, India. pp. 123–128 (Katha), pp. 129–143 Natural Dye, pp. 145–150 (Tannins).

Bawa, R., Gupta, R., Rai, K., Dutt, B., Brahmi, M.K. [Eds.] (2009). *Advance Training on Non–timber Forest Produce*. Published by Dr. Rajan Bawa, Coordinator, Regional Centre, Nature Afforestation and Eco-development Board, Dr. Y.S. Parmar University of Horticulture and Forestry, Nauni, Solan, Himachal Pradesh, India. pp. 180–184.

Dayal, and Thapliyal, M. (2006). Project completion report of project entitled"Chemical Screening of oilseeds, and development of seed handling practices and plantation trial of some high yielding tree species in the Himalayan region (component A and B). Submitted to Department of Biotechnology, CGO Complex, New Delhi.

Dayal, R. (1976). Chemical investigation of some Indian medicinal plants and synthesis of some new prenylated flavonoids. Thesis submitted for the doctor of philosophy degree of University of Delhi, Delhi.

Dayal, R. (1980). Chemical constituents of *Eucalyptus citriodora* (leaves). *Curr. Sci.*, 49: 116.

Dayal, R. and Seshadri, T.R. (1974). Colourless components of the roots of *Artocarpus heterophyllus*- Isolation of a new compound, artoflavanone. *Indian J. Chem.*, 12, 895.

Ernst, R.R and Anderson, W.A. (1966). Application of Fourier transform spectroscopy to magnetic resonance. *Rev. Sci. Insts.*, 37: 93–102.

Finar, I.L. (1975). *Organic Chemistry*, Volume 2, ELBS with Longman, p. 518.

Gunther, Ernest (1948). *The Essential Oils*, Vols. 1–4, D. Van Nostrand Company, Inc. Toronto, New York, London.

Harborne, J.B. (1998). *Phytochemical Methods: A Guide to Modern Techniques of Plant Analysis.* Chapman and Hall, 302.

Harborne, J.B., Mabry, T.J., Mabry, H. (Editors) (1975). *The Flavonoids, Part 1*, Academic Press, New York. San Francisco.

King, Roy W. and Williams Kathryn R. (1990). The Fourier transform in chemistry-NMR, A glossary of NMR terms, *Journal of Chemical Education*, 67(4): A100–A103.

Krishnamurty, H.G. and Seshadri, T.R. (1965). Chemistry of colour tests for chalcones, flavanones and aurones. *Current Science* 24: 681–685.

Mendiratta, A. (2003). Phytochemical examination of *Cephalotaxus harringtonia* var. *harringtonia* needles. Thesis submitted for the doctor of philosophy degree of Forest Research Institute (Deemed University), Dehra Dun, Uttarakhand, India.

Mendiratta (nee) Chugh, A., Dayal, R., Bartley, J.P. (2007). Two new apigenin glycosides from *Cephalotaxus harringtonia* var. *harringtonia. Natural Products Communication*, 2 (11): 1113–1116.

Pew, John C. (1948). A flavonone from Douglas–Fir heartwood. *Journal of American Chemical Society*, 70(9), 3031–3034.

Shinoda, J. (1928). *Journal of Pharmaceutical Society*, 48: 214.

Shukla, Y.N. and Thakur, R.S. (1990). A triterpenoid saponin from *Panax pseudo-ginseng* subsp. *himalaicus* var. *angustifolius, Phytochemistry*, 29(1): 239–241.

Silverstein, Robert M., Webster, Francis X., Kiemle, David J., Bryce, David L. (2014). *Spectrometric Identification of Organic Compounds*, John Wiley and Sons, Inc.

Taketa, Alexandre T.C., Breitmaier, Eberhard, Schenkel, Eloir P. (2004). Triterpenes and triterpenoidal glycosides from the fruits of *Ilex paraguariensis*. *J. Braz. Chem. Soc.* Vol. 15(2) Sao Paulo Mar./Apr. 2004.

Yamaguchi K. (1970). *Spectral Data of Natural Products*, Volume 1, Elsevier Publishing Company; Amsterdam, London, New York.

5

CHAPTER

Chemical Constituents of Some Useful Plants

INTRODUCTION

Forests are essential for the survival of mankind, flora and fauna on the earth. The Indian flora comprises around 45,500 plant species, occurring on a land mass of 3.29 million square kilometres. Forest products obtained from the forests are divided into two groups, namely, wood and non-wood forest products (NWFPs). Wood includes timber and fuel wood, and all other products obtained from forests are covered under NWFPs, which are defined as the goods of biological origin other than wood, derived from forests, other wooded land and trees outside forests. The non-wood forest products (NWFPs) are of immense importance and include medicinal plants, different parts of essential oil bearing plants, oil bearing seeds from different plants, gums, resins, starches, tannins and natural dye materials, edible plants, spices and a number of other products as mentioned in chapter 1. NWFPs are of commercial value. The revenue generation from NWFPs can be increased by developing the value added products from them. Thus, it is of immense importance to carry out the chemical examination of different parts of the plant species for the isolation and characterization of pure chemical compounds from them which either alone or fractions obtained from their extracts may be responsible for their different biological and other properties. Isolation of essential oils from the different parts of the plants, fatty oils from the seeds, tannins, natural dyes and a number of other products obtained from NWFPs are their value added products and are of commercial value.

Plants can be considered chemical factories that have their biosynthetic pathways programmes to synthesize an array of compounds whose structural diversity is often beyond the imagination of even the most imaginative organic chemists. The isolation and characterization of the pure bioactive chemical compounds from plants have become possible with the advent of modern sophisticated chromatography techniques (for isolation) and spectroscopy techniques (for structure elucidation). The separation techniques include column chromatography (CC), thin layer chromatography (TLC), paper chromatography (PC), gas liquid chromatography (GLC), high performance liquid chromatography (HPLC) and high performance thin layer chromatography (HPTLC). The spectroscopic techniques include, nuclear magnetic resonance (NMR) spectroscopy, infrared spectroscopy (IR), ultraviolet spectroscopy (UV), mass spectrometry (MS), gas liquid chromatography-mass spectrometry (GC-MS) and high performance liquid chromatography- mass spectrometry (HPLC-MS). The ability to screen natural products sources for medicinal and other activities has resulted in uncovering of many novel structures-activity relationship producing both new products drugs themselves as well as new leads from rational drug design.

CHEMICAL COMPOUNDS FROM PLANTS

The different plants with a history of medicinal and other properties had been an important incentive for the organic chemists for their chemical examination. However, despite the many important past contributions from the plant kingdom, numerous plant species still remain to be explored for their bioactive principles. Taking this into consideration, the chemical constituents of different parts of the 29 plant species having medicinal and other activities were isolated and characterized using latest separation and spectroscopic techniques. A total of 108 compounds were identified of which 8 were found to be new. The research work on isolation and characterization of chemical constituents, from 29 plant species, was taken during my research tenure either alone or as a team leader/member of the research team and is summarized below.

(A) *Tetrameles nudiflora* R.Br.

1. **Name of plant species**: *Tetrameles nudiflora*
2. **Family**: Datiscaceae, Tetrameliaceae
3. **Popular Name**: Baing-Tree, Maina-Tree
4. **About the plant**: *Tetrameles nudiflora* is a large deciduous tree found in Eastern Himalayas, Assam, Arunachal Pradesh, up to an altitude of about 600 m and also in low level forests of Western ghats, and in the Andamans. It is also dispersed in Sikkim, Sri Lanka and Myanmar. The trees attain the height up to 30–45 m., 9 m. of girth and a clean bole up to 30 m (Anon, 1976, reprinted 2009).
5. **Medicinal and other uses:** The bark of *Tetrameles nudiflora* has been reported to possess laxative and diuretic properties, while its decoction is useful in the treatment of rheumatism, oedema, ascities and icterus. Its wood is not durable and may be used for tea-boxes, match-sticks, ceiling planks and other purposes. It is a good fuel wood having calorific value of 5042 calories (Anon, 1976, reprinted 2009.)

6. **Research contribution of Dayal and Dobhal:**
   (a) **Number of compounds isolated and characterized from bark and leaves:** One
   (b) **Name of compound:** Bryonolic acid (3 *β*-hydroxy-D:C-frideoolean-8-en-29-oic acid)
   (c) **Mode of structure elucidation:** The structure of the compound was elucidated as bryonolic acid by recording and interpretation of its spectral data (IR, ^{1}HNMR, MS) and those of its methyl ester (Dayal and Dobhal, 2001). The identification of the compound as bryonolic acid was further confirmed by comparing the ^{1}HNMR and mass spectra of its methyl ester with methyl bryonolate reported in literature (Sim and Lee, 1972) and by direct comparison with an authentic sample (Kamisako et al., 1984), co-tlc, mmp and superimposable IR spectra. The structure of bryonolic acid is given at the end of this chapter.
   (d) **Pharmacological activity:** Bryonolic acid was found to possess anti- inflammatory and hepatoprotective activities in rats (Dayal and Dobhal, 2001). It also exhibited antiallergic effects in rats (Tanaka et al., 1991).
   (e) **Patent:** A process was developed and patented to isolate bryonolic acid (3*β*-hydroxy-D:C-friedo-olean-8-en-29-oic acid) from the bark (3.0%) and leaves (2.80%) of *Tetrameles nudiflora* (Dayal and Dobhal, 1995).
7. **Conclusion:** The bark and leaves of *Tetrameles nudiflora* may be used for the production of bryonolic acid, which is of medicinal value. The isolation process is very simple.

REFERENCES

Anon. (1976, reprinted 2009). *The Wealth of India: A Dictionary of Indian Raw Materials and Industrial Products. Raw Materials*, Vol. X: Sp-W, pp. 198-199. Council of Scientific and Industrial Research, Publications and Information Directorate, New Delhi, India.

Dayal, R., and Dobhal, P.C. (1995). A process for the preparation of bryonolic acid from *Tetrameles nudiflora* bark and leaves. *Patent* No. 18336 dated 2.6.1995.

Dayal, R. and Dobhal, P.C. (2001). *Tetrameles nudiflora*- A new potential source of bryonolic acid. *Journal of Medicinal and Aromatic Plant Sciences*, 23 (4): 660–661.

Kamisko, W., Suwa, K., Morimoto, K. and Isoi, K. (1984). Application of biosynthetic ^{13}C-enrichment using [1-^{13}C], [1,2-$^{13}C_2$]-acetate as precursor for ^{13}C-NMR spectral assignment of some bryonolic acid derivatives. *Organic Magnetic Resonance*, 22: 93–100.

Sim, K.Y. and Lee, H.T. (1972). Triterpenoid and other constituents from *Sandoricum indicum*. *Phytochemistry*, 11: 3341–3343.

Tanka, Uno C., Akitomo, M., Tabata, M., Honda, C., Kamisako, W. (1991). Anti-allergic effects of bryonolic acid from *Luffa cylindrica* cell suspension. *Planta Medica*, 57: 527–530.

(B) *Vateria indica* Linn.

1. **Name of Plant**: *Vateria indica*
2. **Family:** Dipterocarpaceae
3. **Popular Name:** White Dammar, Indian Copal-Tree
4. **About the plant:** *Vateria* is a small genus of resin yielding trees localized to tropical Asia. Two *Vateria* species namely *Vateria indica* and *Vateria macrocarpa* have been reported to occur in India. *Vateria indica* trees are found in the evergreen forests of western ghats from North Kanara to Kerala in India. It is a tall evergreen tree and normally attains up to 30 m, of height with a clean, cylindrical bole of c 15 m. and a girth of 4.5 m. Its occurrence have been noted in the forests of the lower hills up to an altitude of 750 m; extending to the adjoining area West Coast and Deccan tablelands (Anon, 1976, reprinted 2009).
5. **Medicinal and other uses:** The resin (White Dammar) obtained from the tree and the fat of the seeds known as *Dhupa Fat* are of commercial value. The resin is of medicinal value and is endowed with tonic, carminative and expectorant properties. It is also used for the treatment of a number of ailments, such as throat troubles, chronic bronchitis, piles, diarrhoea, rheumatism, and other diseases. The *Dhupa Fat* is eatable after refining. It has also been used for making candles and soaps. The leaves juice is helpful in curing burns and also for the blood diseases. The bark finds application as an alexipharmic in the Ayurvedic preparations. The timber known as Malabar White Pine is very suitable for durable plywood making (Anon, 1976, reprinted 2009).
6. **Research contribution of Dayal:**
   - (a) **Number of compounds isolated and characterized from the leaves and roots:** Two
   - (b) **Names of compounds:**

     i. Bergenin

     ii. Hopeaphenol
   - (c) **Mode of structures elucidation:** The structures of the two compounds were elucidated as bergenin and hopeaphenol using the spectral data (UV, IR, ^{1}HNMR, ^{13}CNMR and MS). Bergenin was also identified by direct comparison with an authentic sample and by mmp, co-tlc and superimposable IR spectra (Dayal, 1987). The structures of the compounds are given at the end of the chapter.
   - (d) **Pharmacological and other activites:** The aqueous decoction of three herbal plants, viz., *Didymocarpus pedicellata, Bergenia ciliata and Berginia ligulata* has been claimed to be successful for dissolving stone in kidney and blader by

ayurvedic experts. The role of bergenin in this as well as in other drugs is yet unknown (Bharara and Seshadri, 1966). Hopeaphenol was found to possess fungicidal activity against two rotting fungi, namely, *Poria monticola* (at 0.05% and 0.1% level) and *Polyporus versicolor* (at 0.1% level) (Dayal, 1987).

7. **Conclusion:** Bergenin and hopeaphenol were isolated and characterized for the first time from the leaves and roots of *Vateria indica*. The leaves can be used as source of bergenin as it occurs in 1 per cent yield in them.

REFERENCES

Anon. (1976, reprinted 2009). *The Wealth of India: A Dictionary of Indian Raw Materials and Industrial Products. Raw Materials*, Vol. X: Sp-W, pp. 436–441. National Institute of Science Communication and Information Resources, Council of Scientific and Industrial Research, Publications and Information Directorate, New Delhi, India.

Bharara, S.C and Seshadri, T.R. (1966). *Current Science*, **35:** 486.

Dayal, R. (1987). Phytochemical investigation of *Vateria indica* (leaves and roots). *J. Indian Chem. Soc.*, 64: 259.

(C) *Eucalyptus* hybrid (Mainly *E. umbellata* Domin syn. *E. tereticornis* Sm.)

1. **Name of the plant:** *Eucalyptus* hybrid
2. **Family:** Myrtaceae
3. **Popular name:** Safeda
4. **About the plant:** *Eucalyptus* is large genus of evergreen aromatic trees. Its natural occurrence has been observed in Australia and closely adjacent islands, where about 600 species and varieties exist. The easy cultivation of *Eucalyptus* species, their quick growth, thin crown, clean straight bole, aesthetic look and their wide range of adaptability, have led to their widespread introduction into many countries resulting to about 6 million hectares in sixty countries of the world (Dayal, 1994). Eucalyptus plants were first introduced in India in 1790, in the palace garden at Nandi Hills near Mysore, but the regular trials of *Eucalyptus globulus* were started by captain Cotton in 1843 at Wellington in the Nilgiri Hills (FAO, 1979). The Nandi provenance of *Eucalyptus tereticornis* popularly known as *Eucalyptus* hybrid or Mysore gum has been the most widely used species for raising plantations in denuded and barren areas. It is extensively grown in India under the social forestry programme due to its high biomass yield in short span of time. The total area under its plantations was 4.28 lakh hectares in the end

of twentieth century in different states (Anon, 1952, reprinted 2010; Anon, 2002, reprinted 2009; Dayal, 1994).

5. **Medicinal and other uses:** The various *Eucalyptus* species grown in the different countries of the world are of multipurpose use. *Eucalyptus* hybrid wood is mainly used as the raw material for pulp and paper, while the leaves yield an essential oil consisting of cineole, pinenes, terpinene or phellendrene and a number of other compounds (Dayal, 1994). The essential oils isolated from the *Eucalyptus* species leaves may be of perfumery, pharmaceutical and industrial value depending upon the species from which the essential oil is isolated. Besides these uses, the essential oil also exhibit a broad spectrum of biological activity including acaricidal, anti-microbial, fungicidal, insecticidal/insect repellant, herbicidal and nematicidal (Varshney et al., 2012). *Eucalyptus* hybrid wood is being used for making doors and frames and for a number of other purposes. The bark may be used for the preparation of oxalic acid by oxidation process (Prabhu and Theagarajan, 1977).

6. **Research contribution of Dayal et al.:**

   (a) **Number of compounds isolated and characterized from the leaves and wood:** Seven

   (b) **Names of compounds:**

   (i) Ursolic acid

   (ii) Ursolic acid lactone

   (iii) 2-α-hydroxyursolic acid

   (iv) Gallic acid

   (v) 7-β-D-glucosyloxy-5-hydroxy-2-methylchromone

   (vi) (+)-Catechin

   (vii) β-Sitosterol

   (c) **Mode of structure elucidation:** The structures of the above cited seven compounds 6(b) were elucidated by using their spectral data (IR, UV, ^{1}HNMR, ^{13}CNMR) depending upon the complexity of the compound, derivatisation and direct comparison with authentic sample where ever available (Dayal, 1987; Dayal and Dobhal, 1991). The structures of the compounds are given at the end of the chapter.

   (d) **Pharmacological and other uses:** The major pure compound ursolic acid isolated from *Eucalyptus* hybrid leaves was evaluated for its choleretic, anticholestatic and hepatoprotective activities in rats. The data indicated its dose dependent (5–20 mg/kg) choleretic effect. Considerable anticholestic activity (27.9–100%) was noticed against paracetamol (2.0 g/kg) induced cholestasis. Ursolic acid also showed significant hepatoprotective activity against paracetamol and galactosamine (800 mg/kg) induced hepatotoxicity by reversing the altered values in viability of the isolated hepatocytes and the altered biochemical liver and serum parameters. The activity of ursolic acid compared well with already known drug, silymarin (Shukla et al., 1992; Patnaik et al., 1992). Ursolic acid also showed a dose dependent (5–20 mg/kg) hepatoprotective acitivity (21–100%) against thioacetamide and carbon tetrachloride induced hepatotoxicity in rats comparable to silymarin (Saraswat et al., 1996). The essential oil, methanol extract, and ursolic acid showed varied antifungal activity against the three plant pathogenic fungi (*Aspergillus niger, Fusarium solani, Ganoderma lucidum)* and two wood decay fungi (*Pycnoporous sanguineus, Trametes versicolor)* with minimum inhibitory concentrations ranging from 0.5 to 3.0 % (Varshney et al., 2012). Ursolic acid also exhibited significant antifeedant activity against poplar defoliator, *Clostera cupreata* and gave over 92% protection of poplar leaves from its pest (Ahmad et al., 1997). Methods were also developed for the isolation of dye from the leaves and bark of *Eucalyptus* hybrid (Dayal and Dobhal, 2001).

7. **Conclusion:** The *Eucalyptus* hybrid leaves available in plenty may be used for the isolation of essential oil, medicinally important ursolic acid and dye for dyeing the fabrics. The hepatoprotective activity of ursolic acid compared well with the well known drug silymarin and thus have potential for its use as hepatoprotective drug after further studies. Ursolic acid also showed significant antifeedant activity against the poplar defoliator *Clostera cupreata* and antifungal activity against three plant pathogenic fungi and two wood decay fungi as mentioned 6(d) above. The leaves left out after the isolation of essential oil, dye and ursolic acid may be converted to compost, there by leading to the complete utilization of abundantly available *Eucalyptus* hybrid leaves. The bark available in plenty may also be used for isolation of dye and preparation of compost.

REFERENCES

Ahmad, M., Dayal, R., Mishra, R.K. and Dobhal, P.C. (1997). Antifeedant potency of some plants produce against defoliating pest of poplar, *Clostera cupreata. Indian Forester*, 123 (9), 821-826.

Anon. (1952 reprinted 2010). *The Wealth of India: A Dictionary of Indian Raw Materials and Industrial Products. Raw Materials*, Vol. III: D–E, pp. 203–17. National Institute of Science Communication and Information Resources, Council of Scientific and Industrial Research, Publications and Information Directorate, New Delhi, India.

Anon. (2002 reprinted 2009). *The Wealth of India: A Dictionary of Indian Raw Materials and Industrial Products. Raw Materials*, First Supplement Series,Vol. III: D-I, pp. 96–106. National Institute of Science Communication and Information Resources, Council of Scientific and Industrial Research, Publications and Information Directorate, New Delhi, India.

Dayal, R. (1987). Occurrence of ursolic acid and related compounds in *Eucalyptus* hybrid leaves. *Curr. Sci.*, 56: 670.

Dayal, R. (1994). *Eucalyptus hybrid* a renewable source of useful chemicals. *Journal of Non-Timber Forest Products*, 1 (1 & 2): 15–18.

Dayal, R. and Dobhal, P.C. (1991). Chemical constituents of *Eucalyptus* hybrid. *J. Ind. Chem. Soc.*, 68 (8): 478.

Dayal, R. and Dobhal P. C. (2001). Natural dye from Indian plants. *Colourage* Vol., XLVIII (8): 33–38.

FAO, *Eucalyptus for Planting* (1979). Forestry series No. 11, p. 294, Food and Agriculture of the United Nations, Rome

Patnaik, G.K., Shukla, B., Visen, P.K.S., Tripathi, S.C., Dayal, R., Srimal, R. C. (1992). Hepatoprotective activity of ursolic acid isolated from *Eucalyptus tereticornis. Indian J. Pharmcol.*, 24: 47.

Prabhu, V.V and Theagarajan, K.S. (1977). Utilisation of *Eucalyptus* hybrid (Mysore gum) bark for production of oxalic acid. *Indian Forester*, 103 (7): 477–479.

Saraswat, B., Visen, P.K.S., Dayal R., Agarwal, D.P. Patnaik, G.K. (1996) Effect of ursolic acid of chemical induced hepatotoxicity in rats. *Indian J. Pharmcology*, 28: 232–39.

Shukla, B., Visen, P.K.S., Patnaik, G.K., Tripathi, S.C., Srimal, R.C., Dayal, R., Dobhal P.C. (1992). Hepatoprotective activity of ursolic acid isolated from *Eucalyptus* hybrid. *Phytotherapy Research*, 6: 74–79.

Varshney, V.K., Pandy, A., Onial, P.K. and Dayal R. (2012) Antifungal activity of phytochemicals from *Eucalyptus* hybrid leaves against some plant pathogenic and wood decay fungi. *Archives of Phytopathology and Plant Protection*, 45(19): 2347–2354.

(D) *Eucalyptus citriodora* Hook. syn. *E. maculata* Hook. var. *citriodora* Baily

1. **Name of plant species:** *Eucalyptus citriodora*
2. **Family:** Myrtaceae
3. **Popular name:** Lemon-Scented Eucalypt
4. **About the plant:** The necessary details about the genus *Eucalyptus* have been described under *Eucalyptus* hybrid. *Eucalyptus citriodora* is a valuable timber tree being cultivated in Australia, India and many other countries of the world.

It is being grown in the plains of north India and in the sub-montane tract. Its successful trials have been conducted in Malabar, Nilgiris and Coorg. It is generally planted in the gardens in plains due to its smooth trunk and valuable timber (Anon, 1952, reprinted 2010).

5. **Medicinal and other uses:** The *Eucalyptus citriodora* is being grown in many countries for its wood and essential oil from the leaves. The essential oil present in the leaves (0.5–2.0%) finds use in perfumery and as a natural source of citronellal which is being used in the preparation of citronellol, hydroxyl citronellal and menthol. The wood finds use in general construction purposes, carriage building and other uses. The paste of the leaves is applied on the forehead, and a decoction of the leaves in given in cold and cough (Anon, 1952, reprinted 2010; Anon, 2006, reprinted 2010).

6. **Research contribution of Dayal:**

   (a) Number of compounds isolated and characterized from leaves and roots: Seven

   (b) Names of compounds

Compounds From Leaves

(i) Eucalyptin (5-hydroxy-4′, 7-dimethoxy-6,8-dimethylflavone)
(ii) *β*-Sitosterol
(iii) Betulinic acid
(iv) Ursolic acid

Compounds From Roots:

(v) *β*-Sitosterol
(vi) Betulinic acid
(vii) Ursolic acid,
(viii) 7-O-Methylaromadendrin
(ix) Aromadendrin
(x) Fustin

(c) Mode of structure elucidation: The structures of seven pure compounds isolated from leaves and roots as given under 6(b) were elucidated by their colour reactions, physical properties, derivatisation, spectral data (IR, UV, ^{1}HNMR, Mass) and direct comparison with authentic samples (Co-TLC, mmp and superimposable IR) depending on the availability of the sample (Dayal, 1980; Dayal, 1982). The structures of the compounds are given at the end of the chapter.

7. **Conclusion:** Ursolic acid, betulinic acid (from leaves) and fustin (from roots) were reported for the first time in the genus *Eucalyptus* at that time. Aromadendrin and 7-O-methylaromadendrin were isolated for the first time from the roots of an *Eucalyptus* species.

REFERENCES

Anon. (1952 reprinted 2010). *The Wealth of India: A Dictionary of Indian Raw Materials and Industrial Products. Raw Materials*, Vol. III: D-E, pp. 203–217. National Institute of Science Communication and Information Resources, Council of Scientific and Industrial Research, Publications and Information Directorate, New Delhi, India.

Anon. (2006 reprinted 2010). *The Welth of India: A Dictionary of Indian Raw Materials and Industrial Products. Raw Materials*, Second Supplement Vol. I: A-F, pp. 336–346. National Institute of Science Communication and Information Resources, Council of Scientific and Industrial Research, Publications and Information Directorate, New Delhi, India.

Dayal, R. (1980). Chemical constituents of *Eucalyptus citriodora* (leaves). *Curr. Sci.*, 49: 116.

Dayal, R. (1982). Chemical constituents of *Eucalyptus citriodora* (roots). *J. Indian Chem. Soc.*, 49: 1008.

(E) *Eucalyptus melanophloia* F. Muell

1. **Name of plant species:** *Eucalyptus melanophloia*
2. **Family:** Myrtaceae
3. **Popular Name:** Silver-leaved Ironbark
4. **About the plant:** *Eucalyptus* is a large genus of evergreen trees, indigenous to Australia, Tasmania, New Guinea and the neighbouring islands. A large number of *Eucalyptus* species (~100) have been tried in India. *Eucalyptus melanophloia* is a hardy tree which will withstand moderately cold conditions and high temperatures and was first introduced in Eritera (Penfold and Wills, 1961). It was also introduced in the plains of North India (Anon, 1952, reprinted 2010).
5. **Medicinal and other uses:** The leaves are febrifuge, expectorant, disphoretic, antiseptic and stimulant (Nandkarni, 1954), which on steam distillation yield an essential oil consisting mainly of α-phellandrene, α-pinene, p-cymene, cineole, and eudesmol (Anon, 1952, reprinted 2010).

6. **Research contribution of Dayal:**
   (a) **Number of compounds isolated and characterized from leaves:** Five
   (b) **Names of compounds:**
   (i) Eucalyptin (5-hydroxy-4′, 7-dimethoxy-6,8-dimethylflavone)
   (ii) Ursolic acid lactone,
   (iii) Ursolic acid
   (iv) Sideroxylin (5,4′-dihydroxy- 7-methoxy-6,8-dimethylflavone)
   (v) Quercetin-3-O-rhamnoside
   (c) **Mode of structure elucidation:** The structures of the above cited five pure compounds mentioned under 6(b) were determined using colour reactions, derivatisation, hydrolytic studies and spectral data (IR, UV, ^{1}HNMR and mass) and direct comparison with authentic samples (Dayal, 1982). The structures of the compounds are given at the end of the chapter.
7. **Conclusion**: The above cited compounds under 6(b) were reported for the first time from the leaves of *Eucalyptus melanophloia.*

REFERENCES

Anon. (1952 reprinted 2010). *The Wealth of India: A Dictionary of Indian Raw Materials and Industrial Products. Raw Materials*, Vol. III: D–E, pp. 203–217. National Institute of Science Communication and Information Resources, Council of Scientific and Industrial Research, Publications and Information Directorate, New Delhi, India.

Dayal, R. (1982) Phytochemical examination of *Eucalyptus melanophloia* (leaves). *Indian J. Forestry*, **5:** 315.

Nadkarni, K.M. (1954). *India Materia Medica*, p. 514. Popular Book Depot, Bombay.

Penfold, A.R., and Willis, J.L. (1961). *The Eucalyptus, Botany, Cultivation, Chemistry and Utilisation*, p. 266, Inter Science Publishers, Inc; New York.

(F) *Dalbergia stipulacea* Roxb. syn. *Dalbergia ferruginea*

1. **Name of the plant species:** *Dalbergia stipulacea*
2. **Family:** Fabaceae
3. **Popular Name:** East Himalayan *Dalbergia*
4. **About the plant:** *Dalbergia* Linn. is a genus of wood climbers, shrubs and trees extensively dispersed in tropical and sub-tropical geographical areas. The occurrence of about 25 species have been noted in India and *D. stipulacea* is one them. Two species namely *D. sissoo* and *D. latifolia* are the very important Indian timber trees (Anon, 1952, reprinted 2010).

Dalbergia stipulacea occurs specifically in Assam, Arunachal Pradesh, West Bengal and Sikkim of North Eastern India, Chittagong of Bangladesh, Malaysia, Myanmar, Vietnam, West China and in Eastern Himalayas from Nepal to Bhutan (Thothathri, 1987). There are varied reports on the size of its occurrence ranging from a bushy shrubs to a small tree and finally to a large, climbing shrub.

5. **Medicinal and other uses:** The bark and roots may be used as a fish poison (Raizada and Varma, 1937). The wood may be used for making tool handles and posts, while the seeds are reported to be edible. The seeds and leaves may be used as fodder (Anon, 2006, reprinted 2010).

6. **Research contribution of Bhatt, Dayal and Ahmad:**

   (a) **Number of compounds isolated and characterized from leaves and roots:** Three

   (b) **Names of compounds:**

   i. Stipulin (5′,3-di-isopentenyl-2′,4′,4- trihydroxychalcone)

   ii. Luteolin 4′-rutinoside

   iii. Luteolin

   (c) **Mode of structure elucidation:** *Stipulin* isolated from the roots was found to be a new compound and its structure was elucidated by spectral data (IR, UV, ^{1}HNMR, ^{13}CNMR and MS (Bhatt and Dayal, 1992). The synthetic confirmation for the structure of *stipulin* was also provided (Bhatt and Dayal, 2002). The second new compound luteolin 4′- rutinoside was isolated and characterized from the leaves of *Dalbergia stipulacea*. Its structure was elucidated by IR, UV, ^{1}HNMR, ^{13}CNMR, MS and hydrolytic studies (Bhatt and Dayal, 1993). Luteolin was identified by direct comparison with an authentic sample (Co-TLC, mmp and superimposable IR spectra). The structures of compounds are given at the end of the chapter.

   (d) **Pharmacological and pesticidal activities:** Stipulin and luteolin 4′-rutinoside were found to possess antiulcer and mild diuretic activity, respectively in rats. Bark and root extracts of *Dalbergia stipulacea* in alcohol were found to possess significant insecticidal properties on testing against an important defoliator of poplar, *Clostera cupreata* (Ahmad et al., 1996). The acetone extract of bark also showed significant antifeedant activity against the same defoliator (Ahmad et al., 1997).

7. **Conclusion:** Two new compounds namely stipulin from roots and luteolin 4′-rutinoside from leaves were isolated and characterized from *Dalbergia stipulacea*. The leaves can be used as a source of luteolin 4′-rutinoside as it occurs in 1 percent yield in them.

REFERENCES

Ahmad, M., Dayal, R., Misra, R., K. and Dobhal P. C. (1997). Antifeedant potency of some plant produce against defoliating pest of poplar, *Clostera cupreata. Indian Forester*, 123 (9): 821–826.

Ahmad, M., Dayal, R. and Borai P. (1996). Evaluation of insecticidal properties of *Dalbergia stipulacea* against poplar defoliator *Clostera cupreata. Ann. For.*, 4 (1): 47–51.

Anon. (1952 reprinted 2010). *The Wealth of India: A Dictionary of Indian Raw Materials and Industrial Products. Raw Materials*, Vol. III: D–E, pp. 3: 27–29. National Institute of Science Communication and Information Resources, Council of Scientific and Industrial Research, Publications and Information Directorate, New Delhi, India.

Anon. (2006 reprinted 2010). *The Wealth of India: A Dictionary of Indian Raw Materials and Industrial Products. Raw Materials*, Second Supplement Series, Vol. I: A-F, pp. 270–271. National Institute of Science Communication and Information Resources, Council of Scientific and Industrial Research, Publications and Information Directorate, New Delhi, India.

Bhatt, P. and Dayal, R. (1992). Stipulin, Á new prenylated chalcone from *Dalbergia stipulacea. Phytochemistry*, 32 (2): 719.

Bhatt, P. and Dayal, R. (1993). Á flavone glycoside from *Dalbergia stipulacea* leaves. *Phytochemistry*, 33 (3): 731–732.

Bhatt, P. and Dayal, R. (2002), Synthetic confirmation for the structure of stipulin. *Indian J. Chemistry, Sect B,* 41: 184–186.

Raizada, M.B. and Varma, B.S. (1937). Indian plants reputed as fish-poisons. *Indian Forester*, 63(4): 198–218.

Thothathri, K. (1987). *Taxonomic Revision of the Tribe Dalbergieae in the Indian Subcontinent.* Published by Director, Botanical Survey of India, Calcutta.

(G) *Dalbergia sericea* G. Don

1. **Name of Plant:** *Dalbergia sericea*
2. **Family:** Fabaceae
3. **Popular name:** Silky Dalbergia
4. **About the plant:** *Dalbergia sericea* occurs along slopes of mountains at 1400–1600 m, in Arunachal Pradesh, Assam, Meghalaya, Rajasthan (Mount Abu), Sikkim, Uttarakhand, and West Bengal. It is also found in Bhutan Nepal, Myanmar and Vietnam. It is a medium to large-sized tree and attains the height from 40–60 m, having a trunk up to 2 m. with silky velvety young shoots. The tree possesses imparipinate leaves alternate, 14–22 cm long. The tree flowers are exhibited in silky-hairy, dense-flowered panicles, 3–4 cm long, in leaf axils *(www.flowersofindia.net/catalog/slides/silky%20Dalbergia.htmlNov.5,2011; www.ars-grin.gov*/cgi-bin/npgs/html/taxon.pl? 13168*)*.

5. **Medicinal and other uses:** The timber value of *Dalbergia sericea* is not very high because of its comparative lesser durability in comparison with *Dalbergia sissoo* and *Dalbergia latifolia* but used for making very fine wooden articles such as musical instruments (Anon, 1952, reprinted 2010).
6. **Resarch contribution of Dayal, Aiyar, Parthasarthy and Seshadri:**
   (a) **Number of compounds characterized from bark and leaves:** Ten
   (b) **Names of compounds**

Compounds From Bark

(i) Tirucallol acetate
(ii) Glutinol
(iii) β-Sitosterol
(iv) Taraxerol

Compounds From Leaves

(v) ψ-Baptigenin
(vi) β-Sitosterol-β-D-glucopyranoside
(vii) Quercetin
(viii) Quercetin-3-O-galactoside (Hyperin)
(ix) Dihydrokaempferol-3-O-glucoside
(x) Quercetin-3-O- digalactoside

(c) **Mode of structure elucidation:** The structures of the ten pure compounds mentioned under 6(b) were elucidated by spectral data (IR, UV, ^{1}HNMR, MS), hydrolytic studies, derivatisation and by direct comparison with authentic samples depending upon the availability of the samples (mmp, Co-TLC and super impossible IR spectra) [(Aiyar et al., 1974; Dayal and Partasarathy, 1977]. The structures of compounds are given at the end of the chapter.

7. **Conclusion:** This was the first report of the isolation and structure elucidation of the above cited ten compounds mentioned under 6(b), from the bark and leaves of *Dalbergia sericea*. Quercetin-3-O-digalactoside was found to be a new compound.

REFERENCES

Anon. (1952 reprinted 2010). *The Wealth of India: A Dictionary of Indian Raw Materials and Industrial Products. Raw Materials*, Vol. III: D–E, p. 3, 27–29. National Institute of Science Communication and Information Resources, Council of Scientific and Industrial Research, Publications and Information Directorate, New Delhi, India.

Dayal, R. and Parthasarathy, M.R. (1977). Phenolic constituents of *Dalbergiá sericea. Planta Medica*, 31: 245.

Aiyar, V.N., Dayal, R., and Seshadri, T.R. (1974). Triterpenoids of *Dalbergia sericea* (bark). *Curr. Sci.*, 43: 75.

(H) *Adina cordifolia* (Roxb.) Hook. f. ex Brandis

1. **Name of plant:** *Adina cordifolia*
2. **Family:** Rubiaceae
3. **Popular Name:** Haldu
4. **About the plant:** *Adina* is a genus of shrubs and trees, spread throughout the tropics and sub-tropics of Africa and Asia. Five *Adina* species have been reported in India, including *Adina cordifolia*, a useful timber tree. *Adina cordifolia* is found interspread in deciduous forests throughout the greater part of India, ascending to an altitude of 900 m in the sub-Himalayan tract. It is also common in the forest of South India- particularly in the eastern ghats and Karnataka, and attains the height up to 18 m and girth of 6 m (Anon, 1985, reprinted 2010).
5. **Medicinal and other uses:** The bark of *Adina cordifolia* is acrid and bitter, and is used in biliousness. The roots are used as an astringent in dysentery and leaves as fodder. Its heartwood was found to contain a yellow pigment adinin and a number of other compounds alongwith adifoline, corditoline. The timber obtained from *Haldu* tree is used for construction work and for grade1 moisture proof plywood (Anon, 1985, reprinted 2010; Dayal and Dobhal, 1993).
6. **Research contribution of Dayal, Dobhal and Ahmad**
   - **(a) Number of compounds isolated and characterized from leaves:** Two
   - **(b) Names of compounds:**
     - (i) Ursolic acid
     - (ii) Hyperin (Quercetin 3-O-galactoside)
   - **(c) Mode of structure elucidation:** The structure of ursolic acid was elucidated by preparation of derivatives, ^{1}HNMR and direct comparison with authentic sample (m.m.p; Co-TLC and superimpossible IR spectra), while hyperin (Quercetin 3-O-galactoside) was characterized by spectral data (UV, IR, ^{13}CNMR, ^{1}HNMR), and hydrolytic studies (Dayal and Dobhal, 1993). The structures of compounds are given at the end of the chapter.
   - **(d) Pharmacological and other activities:** Ursolic acid is of medicinal value and earlier have been found to possess hepatoprotective activity. The *n*-butanol soluble fraction of the alcohol extract of the leaves exhibited significant insecticidal activity against the larvae of poplar defoliator, *Clostera cupreata* (Ahmad et al., 1996).

7. **Conclusion:** Ursolic acid and hyperin were reported for the first time from *Adina cordifolia* leaves. The *n*-butanol soluble fraction of the alcohol extract of the leaves may provide an eco-friendly way of protecting the poplar plants, after detailed studies.

REFERENCES

Ahmad, M. and Dayal, R. (1996). Evaluation of insecticidal properties of *Dalbergia stipulacea* against poplar defoliator *Clostera cupreata. Ann. For.*, 4 (1): 47–51.

Anon. (1985 reprinted 2010). *The Wealth of India: A Dictionary of Indian Raw Materials and Industrial Products. Raw Materials*, Vol.-I: A, pp. 82–83. National Institute of Science Communication and Information Resources, Council of Scientific and Industrial Research, Publications and Information Directorate, New Delhi, India.

Dayal, R. and Dobhal, P. C. (1993). Chemical components of *Adina cordifolia* leaves. *J. Indian Chem. Soc.*, 183.

(I) *Artocarpus heterophyllus* Lam. syn. *A. integrifolia* Hook. f. (Fl. Br. Ind)

1. **Name of plant:** *Artocarpus heterophyllus*
2. **Family:** Moraceae
3. **Popular name:** Jack Tree or *Kathal*
4. **About the plant:** The genus *Artocarpus* consists of small to large evergreen trees, found in India, Sri Lanka, South China, Malaysia and Solomon Islands. Nine *Artocarpus* species have been reported from India of which two give edible fruits and three are important timber sources. *Artocarpus heterophyllus*, a large evergreen tree up to 10–15 m tall, occurs in the evergreen forests of western ghats at altitudes of 450–1200 m and also grown in the hotter parts of India. It yields edible Jack fruit, very popular in South India and known as poor man's food due to its richness in carbohydrates (Anon, 1985, reprinted 2010).
5. **Medicinal and other uses:** The unripe and ripe fruits are of medicinal value, the former are acrid, astringent and carminative, while the later are laxative, cooling, flattening and useful in biliousness. The leaves are useful in skin diseases, while the seeds are diuretic. The roots find use in asthma, skin disease and diarrhoea. The wood, bark and roots are used in dyeing. The yellow dye isolated from these parts was used for dyeing robes using alum as mordant. The wood of *A. heterophyllus* is fairly hard and not attacked by white ants or fungi. It is widely used for making boxes, furniture, cabinet work, turning, brush-backs and for other purposes. The colour obtained by using the roots of Jack tree was accepted

as the standard in some countries for the robes of Buddhist monks (Chintamani and Subarayappa, 1971).

6. **Research contribution of Dayal and Seshadri:**
   (a) **Number of compounds isolated and characterized from roots:** Nine
   (b) **Name of compounds:**
      (i) Cycloartenone,
      (ii) Artoflavanone
      (iii) Cycloartocarpin
      (iv) Artocarpin
      (v) *β*-Sitosterol
      (vi) Cycloheterophyllin
      (vii) Isocycloheterophyllin
      (viii) Betulinic acid
      (ix) Ursolic acid
   (c) **Mode of structure elucidation:** Artoflavanone (6-C-prenyl-5-hydroxy-7,3′,4′,5′-tetramethoxy flavanone), a new compound, was isolated and characterized from the roots of *Artocarpus heterophyllus* along with eight other known compounds an mentioned in 6(b). The structures of the compounds were elucidated using UV, IR, ^{1}HNMR and MS (mass spectra), derivatisation and direct comparison with authentic samples depending on the availability (Dayal and Seshadri, 1974; Dayal, 1976). Synthetic confirmation for the structure of artoflavanone was also provided, and nineteen new flavonoids were synthesized (Dayal and Seshadri, 1976). The structures of compounds are given at the end of the chapter.
7. **Conclusion:-** *Artoflavanone,* a new compound and eight other compounds were isolated and characterised from the roots of *Artocarpus heterophyllus.* The structures of *artoflavanone* was confirmed by its synthesis. The complete list of compounds isolated by different research group the world over has been summarised in the *Natural Dyes* book (Dayal, 2015).

REFERENCES

Anon. (1985 reprinted 2010). *The Wealth of India: A Dictionary of Indian Raw Materials and Industrial Products. Raw Materials*, Vol.- I:A, p. 444, 447–453. National Institute of Science Communication and Information Resources, Council of Scientific and Industrial Research, Publications and Information Directorate, New Delhi, India.

Chintamani, B.M. and Subbarayappa, B.V. (1971). *Indian J. History of Science*, 6: 102.

Dayal, R. (1976). *Chemical investigation of some Indian medicinal plants and syntheses of some new prenylated flavonoids*, Thesis submitted to University of Delhi, for the Doctor of Philosophy Degree (Ph.D).

Dayal, R. and Seshadri, T. R. (1974). Colourless components of the roots of *Artocarpus heterophyllus*– Isolation of a new compound, artoflavanone. *Indian J. Chem.*, 12: 895.

Dayal, R. and Seshadri, T.R. (1976). Synthesis of artoflavanone and related flavanones. *Indian J. Chem.*, 14: 612.

Dayal, Rameshwar. (2015). *Natural Dyes*. I.K. International Publishing House Pvt. Ltd., New Delhi, p. 186.

(J) *Tectona grandis* Linn. f.

1. **Name of plant:** *Tectona grandis*
2. **Family:** Verbenaceae
3. **Popular name:** Teak, Sagwan
4. **About the plant:** *Tectona* is a small genus of deciduous trees, dispersed in South and South-East Asia and established in other places also. *Tectona grandis* is native to India, Myanmar, Thailand, China and East Indies. In India, it occurs in Madhya Pradesh, extending to parts of Rajasthan, southern Uttar Pradesh, Orissa and in South India. The mechanical properties of Nilambur teak wood from Kerala, Coimbatore teak wood from Tamil Nadu are taken as the standard values for comparison. Its remarkable resistance to termites makes it the most valuable wood for every type of construction. Teak wood finds itself a place wherever strength and durability are important, namely in furniture making, shipbuilding construction of railway carriages and in numerous other uses. This pride of place for teak among timber wood has been recognized for ages, it has been recorded that right in 4000 B.C. the oriental timber traders brought teak wood to Mesopotamia and Yemen, then metropolis of timber trade (Anon, 1976, reprinted 2009; Dayal, 1976).
5. **Medicinal and other uses**: Various parts of the teak have been used in therapy from the distant past. A paste of powdered wood may be used as a remedy for aches and for dispersion of inflammatory swellings. The astringent bark is said to be effective in bronchitis. The oil of the nuts seems to promote the growth of hair and remove itchiness of the skin. The flowers are believed to be useful in biliousness, bronchitis and urinary discharges. Teak leaves are of medicinal value and is a source of dye for dyeing different fabrics. The roots also find use in dyeing of fabrics for mats (Anon, 1976, reprinted 2009; Dayal, 1976).
6. **Research contribution of Dayal and Seshadri:**
   (a) **Number of compounds isolated and characterized from roots:** Eight

(b) Names of compounds:

(i) 1-hydroxy-2-methyl anthraquinone
(ii) Tectoquinone
(iii) Pachybasin
(iv) Dehydrotectol
(v) Tectol
(vi) β-Sitosterol
(vii) Obtusifolin
(viii) Betulinic acid

(c) Mode of structure elucidation: The structures of the compounds mentioned under 6(b) were elucidated by using spectral data (IR, UV, ^{1}HNMR, Mass), derivatisation and direct comparison with authentic samples (Co-TLC, mmp and superimposable IR spectra) depending upon the availability of the authentic samples (Dayal and Seshadri, 1979; Dayal, 1976) The structures of the compounds are at the end of the chapter).

7. **Conclusion:** The chemical examination of the roots of *Tectona grandis* led to the isolation and characterisation of eight compounds, cited under 6(b) for the first time from the roots.

REFERENCES

Anon. (1976 reprinted 2009). *The Wealth of India: A Dictionary of Indian Raw Materials and Industrial Products. Raw Materials*, Vol. X: Sp-W, pp. 136–150. National Institute of Science Communication and Information Resources, Council of Scientific and Industrial Research, Publications and Information Directorate, New Delhi, India.

Dayal, R. (1976). *Chemical investigation of some Indian medicinal plants and syntheses of some new prenylated flavonoids,* Thesis Submitted to University of Delhi, Delhi, for the Doctor of Philosophy Degree (Ph.D.).

Dayal, R. and Seshadri, T.R. (1979). Chemical investigation of *Tectoná grandis* (roots). *J. Indian Chem. Soc.,* 940.

(K) *Gliricidia sepium* (Jacq.) Walp. syn. *G. maculata* (H.B. and K.) Steud.

1. **Name of the plant:** *Gliricidia sepium*
2. **Family:** Leguminosae
3. **Popular name**: Gobbarda Mara (Kannada), Gliricidia (Spanish)

4. **About the plant:** *Gliricidia* H.B. & K, a small genus of shrubs and trees, is the local inhabitant of tropical America and one of its species namely *Gliricidia sepium* is extensively grown in tropics as a shade and decorative tree. *Gliricidia sepium* sometimes known as Madre de Cacao (South America) or Kakawate (Philippines) is a fast growing medium sized, thornless tree which can substitute for *Leuceana leucocephala* as a source of fodder, fuelwood. Its cultivation in India was started in the form of a shade tree in tea, coffee and cocoa plantations. It is a small to medium sized tree having a short bole and cultivated quite extensively in parts of Chennai, Mysore, Mumbai and Travancore-Cochin up to an elevation of about 1000 m. Its flowers are purplish pink or white with large leaves per tree having bright green colour above and pale below. The plant is useful as a source of green manure in the paddy cultivation and suggested to be grown on bunds of fields. The *Gliricidia sepium* plants are fast growing and can be raised by seeds and cuttings (Anon, 1956, reprinted 2009).

5. **Medicinal and other uses:** *Gliricidia sepium* plants are very good source of green material and may yield up to 90–136 kg of green leaves per tree per year on lopping in the fifth year. The flowers are used as vegetable in Central America and Philippines and form a good source of nectar for honeybees. The utilisation of wood can be done for stakes, fences, house posts as a fuel wood and for other purposes (Anon 1956, reprinted 2009).

6. **Research contribution of Dayal**:
   (a) **Number of compounds isolated and characterised from flowers:** Four
   (b) **Names of compounds:**
      (i) Kaempferol-3-O-glucoside (Astragalin)
      (ii) Kaempferol-3-O-galactoside (Trifolin)
      (iii) Kaempferol-3-O-robinoside-7-O-rhamnoside (Robinin)
      (iv) Sucrose
   (c) **Mode of structure elucidation:** The structures of the compounds mentioned under 6(b) were elucidated by their spectral data (IR, UV, ^{1}HNMR, Mass), derivatisation, hydrolytic studies, paper chromatography and direct comparison with authentic samples depending on the availability of the samples (Dayal, 1985). The structures of the compounds are given at the end of the chapter.

7. **Conclusion:** This is the first report of the isolation of four compounds namely astragalin, trifolin, robinin and sucrose from the flowers of *Gliricidia sepium*. The flowers can be used as a source for the isolation of robinin which occurs in 1 per cent yield in them.

REFERENCES

Anon. (1956 reprinted 2009). *The Wealth of India: A Dictionary of Indian Raw Materials and Industrial Products. Raw Materials*, Vol. IV: F-G, p. 137–138. National Institute of Science Communication and Information Resources, Council of Scientific and Industrial Research, Publications and Information Directorate, New Delhi, India.

Dayal, R. (1985). Phytochemical Investigation of the Flowers of *Gliricidia sepium*. *J. Indian Chem Soc.*, 62: 71.

(L) *Jacaranda mimosifolia* Humb. & Bonpl. syn. *J. mimosifolia* D. Don; *J. ovalifolia* R. Br.

1. **Name of the plant:** *Jacaranda mimosifolia*
2. **Family:** Bignoniaceae
3. **Popular name:** Blue *Jacaranda,* Neeli gulmohur (Hindi)
4. **About the plant:** *Jacaranda* Juss., a genus of shrubs and trees, is indigenous to tropical America. Some of its species are being grown in the garden of India as decorative plants. *Jacranda mimosifolia,* a pleasing shrub or medium sized-tree is usually grown in gardens of India for its finely divided foliage and attractive bluish violet flowers. It grows vigorously in well-drained soil and does not accept wet conditions. The plants may be propagated with transplants raised either from seeds or cuttings of half-ripe shoots. The plants withstand pruning and are useful for road side plantations (Anon, 1959, reprinted 2010).
5. **Medicinal and other uses:** The leaves and bark of *Jacaranda mimosifolia* find use in the treatment of syphilis and blennorrhagia in South America. The liquid extract of the leaves is given as a pectoral while the liquid extract of the bark finds use as a lotion for ulcers. The odoriferous wood of the plants is fairly hard which can be used for making tool handles and for other suitable purposes (Anon., 1959, reprinted 2010).
6. **Research contribution of Dayal and Seshadri**:
   - **(a) Number of compounds isolated and characterised from wood:** Four
   - **(b) Names of compounds:**

     (i) Ferulic acid with mixture of C_{19}, C_{21} andC_{23} homologues

     (ii) Spinosterone

     (iii) Caffeic acid ester with C_{23} side chain

     (iv) Ferulic acid with a mixture of C_{28} andC_{29} homologues

(c) **Mode of structure elucidation:** The structures of four compounds isolated from wood as mentioned under 6(b) were identified by spectral data (IR, UV, ^{1}HNMR, and Mass) and hydrolytic studies (Dayal and Seshadri, 1979). The structures of the compounds are given at the end of the chapter.

7. **Conclusion:** This is the first report of the isolation and characterisation of the four compounds mentioned under [6(b)] from the wood.

REFERENCES

Anon. (1959 reprinted 2010). *The Wealth of India: A Dictionary of Indian Raw Materials and Industrial Products. Raw Materials*, Vol. V: H-K, pp. 277–278. National Institute of Science Communication and Information Resources, Council of Scientific and Industrial Research, Publications and Information Directorate, New Delhi, India.

Dayal, R. and Seshadri, T.R. (1979). Chemical examination of *Jacaranda mimosifolia* wood. *J. Indian Chem. Soc.*, 1269.

(M) *Syzygium cumini* (Linn.) Skeels syn. *Eugenia jambolana* Lam.; *cuminii* Druce

1. **Name of the plant species:** *Syzygium cuminii*
2. **Family:** Myrtaceae
3. **Popular name:** Jaman
4. **About the plant:** The genus *Syzygium* Gaertn. includes shrubs and trees dispersed in the tropics of the Old World. The edible fruits are obtained from some of the species belonging to this genus, while a few are sources of good timber and one species provide the cloves of trade. About 75 species grow in India and *Syzygium cumini* is one of them. It is an evergreen tree gaining about 30 m. in height and 3.6 m. in girth, with a trunk length of about 15 m., found all over India up to altitude of 1,800 m. with brown or greyish bark. The moisture containing soil contributes in its growth. The natural propagation of plants is through germination of seeds dispersed by birds and flying foxes which eats the fruits. A distance of 12 m. is kept between two trees on road sides and avenue plantations, and 6 m. for the trees grown for wind–break. It is planted as a shade tree for coffee in Coorg, the Nilgiris, and Kodaikanal (Anon., 1976, reprinted 2009).

5. **Medicinal and other uses:** The pulp, constituting about 75 per cent of the purple colour edible fruit, and eaten widely throughout India. Fruits are also used for preparing squashes, jellies and preserves. The grey to brown seasoned wood is moderately durable in open and very durable under water. The flowers are

important for the honey bees (*Apis dorsata* Fabr.) for making honey in North India. The leaves are used as fodder for cattle, sheep and goats; for green manure and feed for *tasar* silkworms. The seeds are used as feed for livestock. The astringent nature of the bark make it suitable for preparation of gargles and mouth-washes. Bark and seeds extract, and also leaves are used in the treatment of diabetes. The decoction of bark and also powdered seeds finds use in the treatment of diarrhoea and dysentery. The fruits are useful in diabetes. The bark may also be used for dyeing of fabics, fishnets and for tanning purposes (Anon, 1976, reprinted, 2009).

6. **Research contribution of Bhargava, Dayal and Seshadri**:
   (a) **Number of compounds isolated and characterised from bark:** Eleven
   (b) **Names of compounds:**
      (i) Betulinic acid
      (ii) Friedelin
      (iii) Friedelan-3α-ol
      (iv) β-Sitosterol
      (v) Kaempferol
      (vi) Quercetin
      (vii) β-Sitosterol-β-D-glucoside
      (viii) Kaempferol-3-O-glucoside
      (ix) Sucrose
      (x) Gallic acid
      (xi) Ellagic acid
   (c) **Mode of structure elucidation:** The structures of the compounds mentioned under 6(b) were elucidated by using IR, UV, derivatisation, hydrolytic studies and direct comparison with authentic samples using mixed m.p. and superimposable IR spectra (Bhargava et al., 1974). The structures of the compounds are given at the end of the chapter.
7. **Conclusion:** The presence of β-sitosterol–β-D-glucoside, kaempferol-3-O-glucoside, kaempferol and querectin were reported from the bark for the first time. The presence of gallo- and ellagi-tannins in alcohol extract may be responsible for the astringent property of the stem bark.

REFERENCES

Anon. (1976 reprinted 2009). *The Wealth of India: A Dictionary of Indian Raw Materials and Industrial Products. Raw Materials*, Vol. X: Sp-W, pp. 100–104. National Institute of Science Communication and Information Resources, Council of Scientific and Industrial Research, Publications and Information Directorate, New Delhi, India.

Bhargava, K.K., Dayal, R., and Seshadri, T.R. (1974). Chemical constituents of *Eugenia jambolana* (Stem bark). *Curr. Sci.*, 43: 645.

(N) *Pinus roxburghii* Sarg. syn. *P. longifolia* Roxb.

1. **Name of the plant species:** *Pinus roxburghii*
2. **Family:** Pinaceae
3. **Popular name:** Chir pine, Himalayan Long-Leaved Pine
4. **About the plant:** *Pinus* Linn. is a large genus of evergreen resin bearing trees. Pines exist extensively in the temperate areas, in the warm temperate and in sub-tropical countries mainly located in the hills. Five *Pinus* species manifest themselves in India and *Pinus roxburghii* is one of them. A number of exotic pines have also been introduced in India, but not on commercial scale. *Pinus roxburghii*, commonly known as *chir pine*, is a tall tree which occurs in India in the Himalayas from Kashmir to Bhutan and in the Siwalik hills, at an altitude of 450–2400 m. It generally attains the height up to 30 m. and girth of 2.5 m having a cylindrical clean bole of about 12 m. which may vary depending on the nature of soil and other parameters (Anon., 1969, reprinted 2009).
5. **Medicinal and other uses:** The pine trees are tapped for commercially important oleoresin which yields turpentine oil and rosin on distillation in steam jacketed or vacuum stills and both the products have numerous uses (Anon., 1969, reprinted 2009).
6. **Research contribution of Dayal, Bhatt, Dobhal and Ayyar**:
   (a) **Number of compounds isolated and characterised from wax:** Three
   (b) **Names and structures of compounds:** The wax isolated from the pine needles was found to consist of three long chain hydroxy acids as given below.
      (i) 12-hydroxydodecanoic acid (Sabinic acid)
      (ii) 14-hydroxytetradecanoic acid
      (iii) 16-hydroxyhexadecanoic acid (Juniperic acid)

   (c) **Mode of structure elucidation:** The solid wax, isolated from the pine needles by extracting them with chloroform, was hydrolysed with potassium hydroxide in ethylene glycol, to yield a mixture of hydroxy acids (62%). The acids present in the mixture were characterised as 12-hydroxydodecanoic acid (sabinic acid), 14-hydroxytetradecanoic acid, and 16-hydroxyhexadecanoic acid (Juniperic acid), by GC-MS studies of methyl esters obtained by iodination of the mixture of hydroxy acids followed by reduction with

Zn dust in acetic acid and esterification with diazomethane. The acids were converted to macrocyclic lactones by intramolecular cyclisation, under high dilution conditions of the iodoacids with anhydrous potassium carbonate in ethyl methyl ketone or the parent acids with para toluenesulphonic acid in dry benzene. The lactones were characterised as 12, 1-dodecanolide; 14, 1-tetradecanolide; 16,1-hexadecanolide by using spectral data and also by GC-MS studies of the mixture of lactones (Dayal et al., 1989). The lactones are of perfumery value. The structures of the compounds are given at the end of the chapter.

7. **Conclusion:** The wax isolated from the pine needles was found to be an estolide consisting mainly of three acids, namely 12-hydroxydodecanoic acid, 14-hydroxytetradecanoic acid and 16-hydroxyhexadecanoic acid which were converted to macrocyclic lactones used in perfumery with long-lasting musk like odour. This is the first report of the conversion of *Pinus roxburghii* wax to macrocyclic lactones of perfumery value.

REFERENCES

Anon. (1969 reprinted 2009). *The Wealth of India: A Dictionary of Indian Raw Materials and Industrial Products. Raw Materials*, Vol. VIII: Ph-Re, pp. 64–83. National Institute of Science Communication and Information Resources, Council of Scientific and Industrial Research, Publications and Information Directorate, New Delhi, India.

Dayal, R., Bhatt, P., Dobhal, P.C., and Ayyar, K.S. (1989). Perfumery lactones from pine needles (*Pinus roxburghii*) wax. *Indian Perfumer*, 33 (4): 242–245.

(O) *Terminalia citrina* Roxb. ex Flem

1. **Name of plant:** *Terminalia citrina*
2. **Family:** Combretaceae
3. **Popular name:** *Haritaki, Monalis*
4. **About the plant:** *Terminalia* is a large genus which includes mainly deciduous trees seldom woody climbers, and is found in tropical Asia, tropical Africa and tropical America with almost equal distribution in all the regions. The occurrence of about 16 *Terminalia* species have been recorded in India and *Terminalia citrina* is one of them. The trees of the genus yields timber of commercial value. The existence of tannins has been observed in almost all the fruits and bark of all the species (Anon, 1976, reprinted 2009). *Terminalia citrina*, a large deciduous tree, exists in mixed evergreen forests and croped up in the foothills of Himalayas from Nepal eastwards to Assam. Its occurrence in Assam and adjoining states have been reported and up to about 450 m in the Khasi Hills.

5. **Medicinal and other uses**: The fruits find medicinal application similar to those of *Terminalia chebula* and the bark is recorded to own diuretic and cardiotonic properties (Anon, 1976, reprinted, 2009). The wood obtained from the tree is used in construction and other purposes. The barks of some *Terminalia* species are the commercial sources of oxalic acid (Bhatia and Ayyar, 1980).
6. **Research contribution of Dayal and Dobhal:**
   - **(a) Number of compounds isolated and characterized:** Two
   - **(b) Names and structures of compounds:**
     - (i) Arjungenin
     - (ii) Arjunglucoside 1
   - **(c) Mode of structure elucidation:** Two compounds were isolated pure from n-butanol soluble portion of the alcohol extract by column chromatography and their structures were elucidated as arjungenin and arjunglucoside 1 using spectral data (IR, ^{1}HMNR, ^{13}CNMR, Mass) and preparation of derivatives (Dayal and Dobhal, 2001). The structure of the compounds are given at the end of the chapter.
7. **Conclusion:** Literature survery indicated no chemical screening of *Terminalia citrina* bark at that point of time. This was the first report of the occurrence of arjungenin and arjunglucoside1 (Dayal and Dobhal, 2001) in the bark. The bark of *Terminalia citrina* was estimated to contain 5.75% of oxalic acid.

REFERENCES

Anon. (1976 reprinted 2009). *The Wealth of India: A Dictionary of Indian Raw Materials and Industrial Products. Raw Materials*, Vol. X: Sp-W, pp. 177–178. National Institute of Science Communication and Information Resources, Council of Scientific and Industrial Research, Publications and Information Directorate, New Delhi, India.

Bhatia, K. and Ayyar, K.S. (1980). Barks of *Terminalia species*–a new source of oxalic acid. *Indian Forester*, 106: 363–367.

Dayal, R. and Dobhal, P.C. (2001). Chemical investigation of medicinally important *Terminalia citrina* bark. *Journal of Medicinal and Aromatic Plant Sciences*, 23 (3): 409–410.

(P) *Terminalia chebula* Retz

1. **Name of the plant species**: *Terminalia chebula*
2. **Family:** Combretaceae
3. **Popular name:** Harra, Myrobalan

4. **About the plant:** *Terminalia chebula* is a large tree, which grows up to 15–20 m in height and 1.5–2.4 m in girth with a cylindrical bole of 4.0–9.0 m occurs in sub-Himalayan tracts from the Ravi eastward to West Bengal and Assam, ascending up to an altitude of 1500 m in the Himalayas. Its fruits are rich in tannin and known as chebulic or Black Myrobalans in the international commercial market (Anon., 1976, reprinted 2009).

5. **Medicinal and other uses:** The fruits are of medicinal value and its pericarp powder is one of the main ingredients of the edible Ayurvedic formulation "Trifla" and also used in dyeing of fabrics. *T. chebula* is a multiproduct crop par excellence providing a number of ecosystem services (Anon., 1976, reprinted 2009)

6. **Contribution of research team consisting of Onial, Dayal, Rawat and Kumar.**

   (a) **Number of compounds isolated and characterised:** Three

   (b) **Names of compounds:**

   (i) Gallic acid
   (ii) Chebulic acid
   (iii) Chebulinic acid

   (c) **Mode of structure elucidation:** Gallic acid, chebulic acid and chebulinic acid in the fruit pericarps were identified by High Performance Thin Layer Chromatography (HPTLC) [Onial et al., 2013]. The structures of the compounds are given at the end of the chapter.

   (d) **Pharmacological and other activities:** The antifungal activity of the methanol extract of the fruits pericarp was evaluated against pathogenic moulds namely *Fusarium solani* (seedlings), *Aspergillus niger* (seeds), *Aspergillus flavus* (seedling) in the laboratory at different concentrations. The complete protection (100% mortality was noted with methanol extract (3%) against *Aspergillus flavus* and *Fusarium solani* (Onial et al., 2015). Dyeing trials were also carried out using the dye isolated (36.24%) from the pericarp with different mordants. The results indicated that the fruits pericarp may also be used as a source of dye (Onial et al., 2015a). The analysis of fatty oil isolated from the seeds was carried out using GLC of the methyl esters obtained from fatty oil after its hydrolysis followed by methylation (Onial et al., 2014).

7. **Conclusion:** Gallic acid, chebulic acid and chebulinic acid were identified in the fruits pericarps using HPTLC. A method was devised for the estimation of gallic acid in the fruits pericarp using HPTLC. The methanol extract of the fruits pericarp exhibited complete protection against *Aspergillus flavus* and *Fusarium solani* at 3 per cent concentration level. Dyeing trials using the aqueous extract

of the fruits pericarps indicated its use as a source of natural dye for dyeing different fabrics. Fatty acids present in the fatty oil of the seeds were also characterised.

REFERENCES

Anon. (1976 Reprinted 2009). *The Wealth of India: A Dictionary of Indian Raw Materials and Industrial Products. Raw Materials*, Vol. X: Sp-W, pp. 157–176. National Institute of Science Communication and Information Resources, Council of Scientific and Industrial Research, Publications and Information Directorate, New Delhi, India.

Onial, P., Dayal, R., Rawat, M.S.M., Kumar, R (2013). Estimation of gallic acid in *Terminalia chebula* fruits pericarp by HPTLC, *UJPAH*, 11(15).

Onial, P., Dayal, R., Rawat, M.S.M., Kumar, R. (2015a). Utilisation of *Terminalia chebula* fruits pericarp as a source of natural dye for textile applications. *Indian Journal of Natural Products and Resources*, 6(2): 114–121.

Onial, P., Rawat, M.S.M., Dayal, R. (2014). Chemical studies of fatty oil of *Terminalia chebula* seeds kernels, *ISSN Print: 2229–7928, ISSN Online*: 2230–7532.

Onial, P; Pandey, A., Rawat, M.S.M., Dayal, R. (2015). *In-vitro* antifungal assessment of fruits pericarp of *Terminalia chebula* against fungus of forestry origin. *International Journal of Herbal Medicine*, 3(1): 52–55.

(Q) *Cephalotaxus harringtonia* (Knight ex Forbes) Koch var. *harringtonia* syn. C. *drupacea* Sieb. & Zucc. var. *harringtonia* (Knight ex Forbes) Pilger

1. **Name of the plant species:** *Cephalotaxus harringtonia* var. *harringtonia*
2. **Family:** *Cephalotaxaceae*
3. **Popular name:** Harrington Plum Yew
4. About the plant: An evergreen genus *Cephalotaxus,* includes resinous shrubs and trees having densely leafy twigs, occur in eastern Asia. Three *Cephalotaxus* species are found in India. *Cephalotaxus harringtonia* var. *harringtonia* is native to China and Japan and also grows at Forest Research Institute, Dehradun, India. It is small tree up to 10 m in height with expanding or arching branches. Its leaves are dark green, 3–5 cm. long, semi-erect to form V-shaped parting, irregularly pinnate, straight or falcate, often spine-tipped (Anon., 1992).
5. **Medicinal or other uses**: The plant is rich in alkaloids *of Cephalotaxus* and *homoerythrina* series. The alkaloids characterized from needles and roots are

cephalotaxine (major), harringtonine, homoharringtonine, isoharringtonine, deoxyharringtone, desmethylcephalotaxine and cephalotaxinone (Powell et al., 1972; Powell and Milkolajczak, 1973). Harringtonine and homoharringtonine exhibit anticancer activity.

6. **Research contribution of Mendiratta, Dayal and Bartley:**
   - **(a) Number of compounds isolated and characterized from the needles:** Three
   - **(b) Names of compounds:**
     - (i) Apigenin
     - (ii) Apigenin-5-neohesperidoside (Apigenin-5-O-β-D-glucopyranosyl-(2→1)-α-L-rhamnopyranoside)
     - (iii) Apigenin-5-(6″-acetylneohesperidoside) [Apigenin-5-O-β-D-(6″-acetylglucopyranosyl-(2→1)-α-L-rhamnopyranoside].
   - **(c) Mode of structure elucidatation** The structures of three compounds as mentioned under 6(b), isolated from the different extracts of needles, were elucidated using spectral data [IR, UV, ^{1}HNMR, ^{13}CNMR, 2DNMR (HMQC, HMBC), Mass spectra], derivatisation and hydrolytic studies (Mendiratta (Nee Chugh) et al., 2007). The structures of the compounds are given at the end of the chapter.
7. **Conclusion**: Three compounds as given under 6(b) were isolated and characterized from the needles of which apigenin-5-neohesperidoside [apigenin 5-O-β-D-glucopyranosyl-(2→1)-α-L-rhamnopyranoside] and apigenin-5-[6″-acetyl-neohesperidoside] [apigenin-5-O-β-D-(6″-acetylglucopyranosyl-(2→1)-α-L-rhamnopyranoside] were found to be new compounds.

REFERENCES

Anon., (1992). *The Wealth of India: A Dictionary of Indian Raw Materials and Industrial Products. Raw Materials,* Vol. III: Ca-Ci, pp. 437–439. National Institute of Science Communication and Information Resources, Council of Scientific and Industrial Research, Publications and Information Directorate, New Delhi, India.

Mendiratta (nee) Chugh, A., Dayal, R., Bartley, J.P. (2007). Two new apigenin glycosides from *Cephalotaxus harringtonia* var. *harringtonia. Natural Products Communication,* 2 (11): 1113–1116.

Powell, R.G., and Mikolajczak, K.L. (1973). Desmethylceplalotaxinone and its correlation with cephalotaxine. *Phytochemistry,* 12: 2987–2991.

Powell, R.G., Weisleder, D., Smith, C.R. (Jr.). (1972). Antitumour alkaloids from *Cephalotaxus harringtonia* structure and activity. *Journal of Pharmaceutical Sciences*, 61: 1227–1230.

(R) *Vitex negundo* Linn

1. **Name of the plant species:** *Vitex negundo*
2. **Popular name:** Nirgundi
3. **Family:** *Verbenaceae*
4. **About the plant:** *Vitex* is a genus of shrubs, trees, extensively scattered in the tropics and warm temperate geographical areas of both hemispheres. The occurrence of about 14 species have been reported in India. Some timber yielding *Vitex* species are of commercial value while others are used in medicine. *Vitex negundo*, an aromatic shrub up to 4.5 m in height occurs throught the greater part of India extending up to an altitude of 1500 m in the outer Himalayas. It is abundant along the bank of rivers, in moist places, open waterlands and near deciduous forests and also extensively planted as a hedge plant along roads, between fields as it is normally not browsed by cattle. It can be produced readily by cuttings and is very useful for planting against soil erosion and for afforestation in the reclamation of flooded forestlands. The leaves are usually with five leaflets (rarely three) which are palmately arranged. The leaflets are lanceolate, 4 to 10 cm long, hairly beneath and pointed at both ends. The flowers are bluish purple. The fruit is succulent drupe, black when ripe, rounded and about 4 mm in diameter (Anon., 1976, reprinted 2009; Chopra et al., 1956; Kirtikar and Basu, 1980). It also occurs in Sri Lanka, Myanmar, China, Pakistan, Afghanistan, Malaysia, Tropical Africa, and Philippines.
5. **Medicinal and Other uses:** All parts of the plants are reported to have medicinal properties and used in indigenous system of medicine. The flowers are astringent and used in fever, diarrhoea and liver complaints. The leaves and roots are more important and sold as drugs and used for various ailments (Anon., 1976, reprinted 2009). The fruits are also of medicinal value and recommended in headache, catarrh, and watery eyes. These are prominently used in China. The boiled seeds are described to be eaten in Philippines. The leaves have been reported to possess broad spectrum pesticidal activity (Anon., 1976, reprinted 2009; Grainge et al., 1985; David et al., 1988; Krishnarajah et al., 1985).
6. **Research contribution of Singh, Dayal, Bartley and Bhandari:**
   (a) **Number of compounds isolated and characterized from the leaves:** 12

Krishnarajah, S.R., Ganeshaligam, U.K., Senanayaka, U.M. (1985). Repellency and toxicity of some plant oils and their terpenes. *Tropical Science* 25(4): 52.

Singh, V. (2004). Phytochemical Investigation of *Vitex negundo leaves.* Thesis submitted to Forest Research Institute, Dehradun for the Doctor of Philosophy (Ph.D.) degree.

Singh,V., Dayal, R., Bhandari, R.S. (2002). Pesticidal activity of essential of *Vitex negundo* leaves against *Sitotroga cerealella*. *Shashpa* 9(1): 71–75.

Singh, V., Dayal, R., Bartley, J.P. (2003). Chemical Constituents of *Vitex negundo* Leaves. *Journal Medicinal and Aromatic Plant Sciences*, 5 (1): 94–98.

(S) *Shorea robusta* Gaertin. f.

1. **Name of the plant species:** *Shorea robusta*
2. **Family:** Dipterocarpaceae
3. **Popular name:** Sal
4. **About the plant:** *Shorea* Roxb. a big genus of resiniferous trees extensively disseminated in India Ceylon, Myanmar and other South-East Asian nations up to Philippine Islands. Four species have been reported from India and *Shorea robusta* is one of them. *Shorea robusta*, normally known as "Sal" is the most plentiful available timber producing tree in India. Sal, a large sub-deciduous tree and seldom quite leafless, attains the height of 18–39 m and a girth of 1.8–2.7 m under average conditions. It is grown widely in parts of North, East and Central India occupying 13.30 per cent of the total forest cover of the country. Bark of the tree is reddish brown or grey in colour and smooth or longitudinally fissured. Leaves are ovate, oblong, and coriaceous, and shining when mature. Flowers are yellowish and small. The plant bears ovoid and reddish to pale yellowish green coloured fruits which are indehiscent and one seeded with five wing-like persistent sepals (Anon., 1972, reprinted 2009).
5. **Medicinal and other uses:** It is a multipurpose tree as its bark, resin, leaves and the seeds available in plently, are used for different purposes including medicinal uses (Anon., 1972 reprinted 2009; Nadkarni and Nadkarni, 1988). The bark is rich in tannin content. The tannin extracted from the bark may be used as a cheap tanning material. Sal trees on tapping yield an oleoresin usually called as Sal Dammar. It is produced in huge quantity in India which has numerous uses, including a component of *samagri* which is burnt in religious functions and cremation rites. It is also of medicinal value and is useful in diarrhoea and dysentery. An essential oil *(Chua oil)* is obtained on the dry distillation of the resin which also have a number of uses. Sal Butter is obtained from the sal seeds which is locally used for cooking, lighting and as a substitute of cocoa butter in the

manufacture of chocolates. Fruits paste is used in diarrhoea. The timber obtained from *Sal* is very hard, heavy, durable and is one of the high cost timber being used in construction of houses and various other uses (Anon., 1972, reprinted 2009).

6. **Research contribution of Sandeep, Varshney, and Dayal:**
   (a) **Number of compounds isolated and characterized from bast, leaves and roots:** Six
   (b) **Names of compounds:**
      (i) β-Sitosterol (bast and roots)
      (ii) β-Sitosterol-β-*D*-glucoside (leaves, roots and bast)
      (iii) Resveratrol (roots)
      (iv) Hopeaphenol (roots and bast)
      (v) Friedelin (roots)
      (vi) Bergenin (leaves and roots)
   (c) **Mode of structure elucidation:** The structures of pure compounds were elucidated by recording and interpretation of their spectral data (UV, IR, ^{1}HNMR, ^{13}CNMR and MS) and also by direct comparison with authentic samples (Co-TLC, mmp and superimposable IR spectra) depending upon their availability (Kaur, 2003); Varsheny and Dayal, 2006, 2009. The structures of the compounds are given at the end of the chapter.
7. **Conclusion:** Six compounds mentioned under 6(b) were identified from the different parts namely bast (cambium + secondary phloem), leaves and roots. Resveratrol was reported for the first time in the *Shorea* genus.

REFERENCES

Anon. (1972 Reprinted 2009). *The Wealth of India: A Dictionary of Indian Raw Materials and Industrial Products. Raw Materials*, Vol. IX: Rh-So, pp. 312–321. National Institute of Science Communication and Information Resources, Council of Scientific and Industrial Research, Publications and Information Directorate, New Delhi, India.

Kaur, S. (2003). Phytochemical examination of *Shorea robusta*, PhD thesis. Forest Research Institute University (ICFRE) Dehradun, India p. 151-182.

Nakarni, K.M., and Nadkarni, A.N. (1988). *Indian Materica Medica*, Popular Prakashan, Bombay, p. 1132.

Varshney, V.K. and Dayal, R. (2006). Chemical constituents of *Shorea robusta, Int. J. Sci.*, 4(2): 298–304.

Varshney, V. K. and Dayal, R. (2009). Phytochemicals from *Shorea robusta*–A Review. Proceedings of the *Conference on Natural Products and Biodiversity: Chemistry and Utilization*, Chemistry Division, Forest Research Institute, Dehradun, 68–75.

(T) *Achyranthes aspera* Linn

1. **Name of the plant species:** *Achyranthes aspera*
2. **Family:** Amaranthacae
3. **Popular name:** *Chirchitta*, Latjira, Prickly-chaff Flower
4. **About the plant:** *Achyranthes* Linn. is a genus of small herbs or small shrubs scattered throughout the tropical and sub-tropical areas. Around nine species exist in India. *Achyranthes aspera*, an upright or procumbent annual or perennial herb of 1–2 m in height occurs throughout India, up to an altitude of 2,100 m and in the South Andaman Islands. It normally occurs as weed, on waysides and unusable land usually, with woody base. It also commonly grows in the shady places of the cultivated fields and *chir* pine plantations as a weed. It tolerate dry conditions, but waterlogging in not favourable for its growth (Anon., 1985, reprinted 2010).
5. **Medicinal and other uses:** The plant is very useful in the indigenous systems of medicine and have numerous uses (Rashmi and Dayal, 2007). It is described to be pungent astringent, pectoral, diuretic and used as an emmenagogue, in piles and haemorrhage and profuse mucous discharge and in numerous other uses (Anon, 1985, reprinted 2010; Anon, 2006, reprinted 2010).

   According to the traditional knowledge of tribals of Patalkot (Madhya Pradesh), Dang (Gujarat), Aravelli (Rajasthan) as collected by Director, Abhumaka Private Limited, Ahmedabad, *Latjira* may find use by the tribals for the following useful purposes:

   (i) The decoction of some seeds in water may be useful for the liver problem.
   (ii) Leaves juice is useful for cold and cough.
   (iii) The seeds on roasting in earthen container may be taken to reduce hunger.
   (iv) The leaves juice helps in stopping the blood flow from fresh wounds or cuts.
   (v) The juice of whole plant is useful in sleeplessness and also helps in sleeplessness due to tension, tiredness and irritation.
   (vi) The chewing of the leaves with asafoetida (hing) gives relief in toothache and also stops the blood from teeth.
   (vii) The stem may be is used as "Datun" to keep the teeth and gums strong and in good health, while the roots are used in the pain in tooth and gums.

6. **Research contribution of Rashmi, Dayal, and Nagatsu:**

   (a) Number of compounds isolated and characterized: 3

(b) Names of compounds:

(i) α-L—rhamnopyranosyl-(1→4)-(β-D-glucopyranosyluronic acid)-(1→3)-oleanolic acid

(ii) α-L—rhamnopyranosyl-(1→4)-(β-D-glucopyranosyluronic acid)-(1-→3)-oleanolic acid-28-O-β-D-glucopyranoside

(iii) α-L—rhamnopyranosyl-(1→4)-(β-D-glucopyranosyluronic acid)-(1-→3)-oleanolic acid-28-O-β-D-glucopyranosyl-(1→4)-β-D-glucopyranoside

(c) Mode of structure elucidation: The structures of the pure isolated compounds mentioned under 6(b) were elucidated on the basis of spectral data, viz., IR, UV, ^{1}HNMR, ^{13}CNMR, 2DNMR (HMBC, HMQC and COSY) and hydrolytic studies. The structures of the compounds are given at the end of the chapter. (Rashmi et al., 2007; Rashmi, 2005)

(d) Pharmacological and other activity: Petroleum ether and acetone extracts of leaves, stem, roots and methanol extract of the seeds exhibited significant antioxidant activity (Rashmi et al., 2007a).

7. **Conclusion**: Two new oleanolic acid glycosides (i) and (iii) along with a known oleanolic acid glycoside (ii) as mentioned under 6(b) above were isolated and characterized from the seeds of *Achyranthes aspera*. The extracts of different parts as mentioned under 6(d) exhibited antioxidant activity.

REFERENCES

Anon. (2006 reprinted 2010). *The Wealth of India: A Dictionary of Inidian Raw materials and Industrial Products. Raw materials*, second supplements Series Vol. I,: A-F, pp. 13-14. National Institute of Science Communication and Information Resources Council of Science and Industrial Research, Publication Information Directorate New, Delhi, India.

Anon. (1985 reprinted 2010). *The Wealth of India: A Dictionary of Indian Raw Materials and Industrial Products. Raw Materials*, Vol. I:A, pp. 55–57. National Institute of Science Communication and Information Resources, Council of Scientific and Industrial Research, Publications and Information Directorate, New Delhi, India.

Rashmi (2005). Phytochemical examination of *Achyranthes aspera*, PhD thesis. Forest Research Institute (ICFRE) Dehra Dun, India p. 97–137.

Rashmi, Dayal, R. and Nagatsu A. (2007). Three oleanolic acid glycosides from the seeds of *Achyranthes aspera. Natural Products Communications,* Vol **2** (7): 727–730.

Rashmi, Dayal, R. and Subramaniam, V. (2007a). *In vitro* antioxidant activity of medicinally important *Achyranthes aspera*–A preliminary study, *Indian Drugs,* 44 (2): 128–131.

(U) *Tagetes minuta* Linn. syn. *T. glandulifera* Schrank

1. **Name of the plant:** *Tagetes minuta*
2. **Family:** Asteraceae, Compositae
3. **Popular name:** Wild marigold, Stinking Roger
4. **About the plant:** *Tagetes* Linn. is a large genus consisting of 51 plant species and *Tagetes minuta* is one of them. It is native of South America and has naturalised in India near waste places and on dry areas in the North-West Himalayas between altitudes of 1250–2500 meters, normally attaining the height of 1–2 metres. It is largely aromatic in nature and used for the isolation of essential oil known as 'Tagetes oil'. Its leaves are 7–15 cm long, with pale yellow flower heads in corymbose clusters (Anon., 1976, reprinted, 2009). The essential oil is of commercial value due to its use in perfumery and flavour industry.
5. **Medicinal and other uses:** The plant is reported to be of medicinal and aromatic value. The oil exhibited tranquilizing, hypotensive, bronchodilatory, spasmolytic and anti-inflammatory properties in standard assays (Chandhoke et al., 1969). The essential oil also showed mosquito larvicidal (Green et al., 1993); antiviral and antifungal activities (Bii et al., 2000; Singh et al., 2002; Tereschuk et al., 1997 and Zygadlo et al., 1994). Decoction of the plant is prepared by extracting the whole herb with water which is used as beverages and a remedy for cold, digestive system complaints, stomach upsets, diarrhoea and live ailments (Singh et al., 2003). The flower oil and its absolute are used for high grade perfumes in flavour beverages, frozen diary deserts, candy baked goods, gelatins, puddings condiments, etc. (Singh et al., 2003).
6. **Research contribution of Kumar, Dayal and Onial.**
   - **(a) Number of compounds isolated and characterised from flowering twigs:** Two
   - **(b) Names of compounds:**
     - (i) β–Sitosterol
     - (ii) Quercetagetin 3-O-β-glucoside
   - **(c) Mode of structure elucidation:** The structures of the above cited [6(b)] compounds were elucidated using spectral data (IR, UV, ^{1}HNMR, ^{13}CNMR, and Mass), hydrolytic studies and direct comparison with authentic samples (Co-TLC, mmp and superimposable IR spectra) [Kumar, 2010]. The structures of the compounds are given at the end of the chapter.
7. **Conclusion:** Two compounds, namely, β-sitosterol and quercetagetin 3-O-β-glucoside have been reported for the first time from the flowering twigs of *Tagetes minuta*.

REFERENCES

Anon. (1976 reprinted 2009). *The Wealth of India: A Dictionary of Indian Raw Materials and Industrial Products. Raw Materials*, Vol. X:Sp-W, pp. 109–112. National Institute of Science Communication and Information Resources, Council of Scientific and Industrial Research, Publications and Information Directorate, New Delhi, India.

Bii, C.C; Siboe G.M. Mibey, R.K. (2000). Plant essential oils with promising antifungal activity. *East African Medical Journal.* 77(6): 319–22.

Chandhoke, N. and Ghatak, B. J. Ray. (1969). *Tagetes minuta*: pharmacological actions of the essential oil. *Indian Journal of Medical Research.* 57(5): 864–76.

Green, M.M., Singer, J.M., Sutherland, D.J. Hibben, C.R. (1991). Larvicidal activity of *Tagetes minuta* (marigold) toward *Aedes aegypti. Journal of the American Mosquito Control Association.* 7(2): 282–6.

Kumar, Rakesh (2010). Chemical examination and study on natural dye from *Tagetes minuta*, Ph.D. thesis, Forest Resarch Institute University (ICFRE), Dehradun, India.

Singh, B., Joshi, V. P., Ram, R., Sharma, A. Zaidi, A. A. (2002). Use of *Tagetes minuta* oil and its components as antiviral agents. (Council of Scientific and Industrial Research, India). U.S. Patent: 17 pp. CODEN: USXXAM US 6444458 B1 20020903

Singh, V., Singh, B., and Kaul Vijay K. (2003). Domestication of wild marigold (*Tagetes minata*) as a potential economic crop in western Himalayas and north India plains. *Economic Botany* 57(4): 535–544.

Tereschuk, Maria L., Riera, Marta V.Q., Castro, Guillermo R., Abdala, L. R. (1997). Antimicrobial activity of flavonoids from leaves of *Tagetes minuta. Journal of Ethno Pharmacology.* 56(3): 227–232.

Zygadlo, J. A., Guzman, C. A. and Grosso, N. R. (1994). Antifungal properties of the leaf oils of *Tagetes minuta* L. and *T. filifolia* Lag. *Journal of Essential Oil Research.* 6(6): 617–21.

(V) *Clematis roylei* Rehder syn. *C. nutans* Royle

1. **Name of the plant species:** *Clematis roylei*
2. **Family:** Ranunculaceae
3. **Popular name:** Nuk-chhikni
4. **About the plant:** *Clematis* is a large genus of about 280 species of climbing plants distributed through out the world. The genus is known for its medicinal activites and it is being used in many traditional systems of medicine. About 25 species occur in India in the temperate parts, which are of medicinal value (Singh et al., 2009). A number of species are acrid and poisonous. *Clematis roylei* is commonly known as Nuk-chhikni or Kangulii in Garhwal (Uttarakhand, India). It

is distributed along the outer range of hills between 2000–5000 feet. The plant is a climber with stem 2–6 feet high and about 0.4 inches in diameter. Twigs ribbed silky-pubescent. Leaves 4–6 inches long pinnate, leaflets 1–3 inches long, ovate or lanceolate, undivided or 3–5 lobed or partite, irregularly toothed, rarely entire, silky when young, ultimately becoming finely ad pressed-hairy on both surface. Flowers pale-greenish yellow, about 0.5 inches long , drooping, in leaf axillary pubescent panicles 2–6 inches long. Sepals pale yellow, erect with recurved tips, 0.7–0.9 inch long, thick, densely silky outside, and glabrous inside (Anon., 1981; Osmaston, 1927; Babu, 1977).

5. **Medicinal and other uses:** *Clematis roylei* is used as an anthelmintic, expectorant in blood and skin disorders, in eye troubles, in hepatic disorders and by inhaling of powdered leaves relieves headaches (Singh et al., 2009). The roots of *C. chinensis* are used in anti-inflammatory conditions in the traditional Chinese medicine (Shao et al., 1995), while *C. glucinoides, C. microphylla, and C. pickeringii, etc.* are used in Australian traditional system of medicine (Singh, 2007).

6. **Research contribution of Singh, Gaidhani and Dayal**
   - **(a) Numbers of compounds isolated characterised from the leaves:** Six
   - **(b) Names of compounds:**
     - (i) β-Sitosterol,
     - (ii) β-Sitosterol-β-D-glucoside
     - (iii) α-Amyrin
     - (iv) Campesterol
     - (v) Stigmasterol
     - (vi) Oleanolic acid
   - **(c) Mode of structure elucidation:** The structures of the compounds mentioned under 6(b) were elucidated by recording their spectral data IR, ^{1}HNMR, GC-MS, derivatisation and direct comparison with authentic samples depending upon their availability (Singh, 2007). The structures of the compounds are given at the end of the chapter.
   - **(d) Phamacological activities:** The alcohol extract of the aerial parts was evaluated for anti-inflammatory activity by carrageenan-induced rat paw edema. The extract produced a significant ($p<0.001$) and dose dependent inhibition of the carrageenan-induced edema (Singh et al., 2009).

7. **Conclusion:** This is the first report of the isolation and characterisation of the compounds from *Clematis roylei* cited under 6(b). The presence of these compounds along with essential oil in the leaves may be responsible for their medicinal properties and also those exhibited by their extracts.

REFERENCES

Anon. (1981). *The Wealth of India: A Dictionary of Indian Raw Materials and Industrial Products. Raw Materials*, Vol. II, pp. 101–103, 124–125. National Institute of Science Communication and Information Resources, Council of Scientific and Industrial Research, Publications and Information Directorate, New Delhi, India.

Babu, C.R. (1977). Herbaceous flora of Dehradun, Publication and Information Directorate (CSIR), New Delhi p. 40–41.

Osmaston, A.E. (1927). A forest flora of Kumaon, Govt. press, United Provinces, Allahabad, p. 6.

Shao, B.P.; Quin, G.; Xu, R; Wu, H; Ma, K. (1995). Triterpenoid saponins from *Clematis Chinensis*. *Phytochemistry*, 38(6): 1473–79.

Singh, R. (2007). Phytochemical examination of *Clematis roylei* Reheder, Ph.D. thesis, Forest Resarch Institute University (ICFRE), Dehradun, India.

Singh, R., Gaidhani, S., and Dayal, R. (2009). Evaluation of anti-innflammatory properties of *Clematis roylei*. *Journal of Herbal Medicine and Toxicology*, 3(2): 165–167.

(W) *Peucedanum grande* C.B. Clarke

1. **Name of the plant species:** *Peucedanum grande*
2. **Family:** Umbelliferae, Apiaceae
3. **Popular name:** Duku, Baphali or Wild Carrot
4. **About the plant:** *Peucedanum* is a genus of herbs, shrubs and seldom trees dispersed namely in Asia, Europe, North-East Africa and Western South America. Ten species exist in India and *Peucedanum grande* is one of them. It is a succulent herb and usually attains the height of about a meter or more, existing gregariously on the Western ghats and hills of Deccan Plateau with large roots and perennial in nature. The stem is fistular and emits a strong scent on crushing. Leaves pinnate or bipinnate, mostly radical; flowers yellow in compound umbels. The reddish yellow fruits are ovate, broadly elliptical with the length of 10–13 mm (Anon., 1966, reprinted 2010).
5. **Medicinal and other uses:** The fruits possessing strong lemon-like odour, are of medicinal value having carminative, diuretic, stimulant and tonic properties. A fruit infusion is useful against gastric and intestinal problems (Anon, 1966, reprinted 2010). The fruits are added to curries as flavouring agent (Nadkarni, 2002). The fruits also possess aphrodisiac, anthelmintic, antifatulent, anti-inflammatory, deobstruent, diaphoretic, and emmenagogue properties (Aslam et al., 2012).
6. **Research contribution of Aslam et al.:**
   - **(a) Number of compounds isolated and characterised from fruits:** Four
   - **(b) Names of compounds:**

(i) 5-hydroxy-6-isopranyl coumarin

(ii) 5, 6-furanocoumarin

(iii) 7-methoxy-5, 6-furanocoumarin

(iv) Labdanyl-3α-ol-18-(3‴-methoxy-2‴-naphthyl-oate)-3α-L-arabinofuranosyl-(2′-1″)-α-L-arabinofuranoside

(c) **Mode of structure elucidation:** The structures of above cited compounds under 6(b) were elucidated on the basis of spectral data (FT-IR, UV, ^{1}HNMR, ^{13}CNMR, FAB mass spectra) (Aslam et al., 2012a). The structures of the compounds are given at the end of the chapter.

(d) **Pharmacological activites:** The methanol extract of *P. grande* fruits exhibited nephroprotective activity against the cadmium chloride and mercuric chloride induced renal toxicity in Wistar rats (Aslam et al., 2012b; Aslam et al., 2012c). The methanol extract and compound (iv) under (6b) showed nephroprotective activity against gentamicin induced nephrotoxicity in Wistar rats (Aslam et al., 2012a).

7. **Conclusion:** The fruits of *Peucedanum grande* are of medicinal value. Four compounds as mentioned under 6(b) were isolated and characterised from the methanol extract of the fruits out of which compound (iv) identified as labdanyl-3α-ol-18-(3‴-methoxy-2‴-naphthyl-oate)-3α-L-arabinofuranosyl-(2′→1″)-α-L-arabinofuranoside (4) was found to be new. The methanolic extract of *P. grande* and compound (iv) exhibited nephroprotective activity against gentamicin-induced nephrotoxicity in Wistar rats.

REFERENCES

Anon. (1966 reprinted 2010). *The Wealth of India: A Dictionary of Indian Raw Materials and Industrial Products. Raw Materials*, Vol. VII:N-Pe, p. 330. National Institute of Science Communication and Information Resources, Council of Scientific and Industrial Research, Publications and Information Directorate, New Delhi, India.

Aslam, M., Ahmad, S.T., Asiaf, A., Javed, K., Dayal, R., Singh, S. (2012c). *Peucedanum grande* attenuates acute renal failure and oxidative stress induced by mercuric chloride in rodents. *Am. J. Pharmtech Re*search, 2(3): 771–782.

Aslam, M., Ahmad, S.T., Dayal, R., Javed, K., Umar S., Asiaf, A., Nafees, Bhat, J.U., Wani, A., Sanin, M., Singh, S. (2012b). Nephroprotective action of *Peucedanum grande* against cadium chloride induced renal toxicity in wistar rats. *EXCLI Journal*, 11: 444–452.

Aslam, M., Dayal R., Javed, K., Fahamiya, N., Mujeeb M., Husain, A. (2012). Development of quality standards and phytochemical analysis of *Peucedaumm grande*. *American Journal of Pharm. Tech Research*, 2(1), 537-548.

Aslam, M. Mohammed, A., Dayal R., Javed, K. (2012a). Coumarins and naphthyl labdanoate diarabinoside from the fruits of *Peucedanum grande* C.B. Clarke. Z. *Naturforch*; 67c: 580–586.

Nadkarni, K.M. (2002). *Indian Materia Medica*. Popular Prakashan Pvt. Ltd., Mumbai p. 934.

(X) *Rheum emodi* Wall. ex. Meissn

1. **Name of the plant species**: *Rheum emodi*
2. **Family:** Polygonaceae
3. **Popular name**: Himalayan Rhubarb, Indian Rhubarb
4. **About the plant:** *Rheum* Linn. is a big genus of perennial, sound herbs, strewed in the temperate and sub-tropical areas of the world, mainly in Asia. A number of species are used as medicine, some for cooking purposes and some are raised as ornamental plants. It is a source of well known drug *Rhubarb* obtained from the dried rhizomes and roots of some *Rheum* species and find use in purgatives, laxatives, and stomach–tonics. Rhubarbs are mostly of three types, namely, the Indian or Himalayan Rhubarb, the Chinese Rhubarb, and Rhapontic Rhubarb comprising of the rhizomes and roots of different *Rheum* species and consist of anthroquinone derivatives (Anon., 1972, reprinted 2009).

   The plant of *Rheum emodi* is a perennial sturdy herb normally 1.5–3.0 m. tall and grows wild in the Himalayas from Kashmir to Sikkim at altitudes of 3,300–5,200 m. It is also grown in Assam for the utilisation of its leaves as vegetable. The dark–purple or pale–red flowers in tall axillary panicles, are small in size. The plants have very strong roots with usually long leaves of about 60 cm in diameter. The plant occurs wild in Himachal Pradesh, Uttrakhand, Nepal and Sikkim from where rhizomes may be collected. This herb may be grown in Himachal Pradesh and Kashmir at an altitude of about 2000 m. The drought-resistant herb is usually propagated through rhizome-cuttings in early spring and the cuttings are planted at a spacing of 1.2–1.5 m by keeping about 10 cm of crown under the surface. The harvesting is normally done in September from 3–10 year old plants (Anon., 1972, reprinted, 2009).

5. **Medicinal and other uses**: *Rheum emodi* (Indian rhubarb) finds use as a purgative astringent tonic, while its stimulating effect in combination with aperient properties, makes it specially useful in atonic dyspepsia. The patients suffering

from gout, rheumatism, epilepsy and uric acid problems are not allowed to use it. The roots are useful in dyeing fabrics, cleaning teeth and quick healing of ulcers. The leaves and flowers are edible (Anon., 1972, reprinted 2009). A number of other medicinal properties like anti-inflammatory, diuretic, emmenogouge, expectorant, hepatoprotective and other uses have been reported in Unani literature (Aslam, 2011).

6. **Research contribution of Aslam et al.**
   (a) **Number of compounds isolated and characterised from rhizomes:** Two
   (b) **Names of compounds:**
      (i) 8-Dehydroxychrysophanol
      (ii) Emodin
   (c) **Mode of structure elucidation:** The structure of 8-dehydroxychrysophanol was elucidated using spectral data (IR, UV ^{1}HNMR, and FAB-MS) and emodin was estimated in methanol extract using HPTLC (Aslam, 2011; Aslam, 2012). The structures of the compounds are given at the end of the chapter.
   (d) **Pharmacological activity:** Petroleum ether and chloroform extracts, and fractions of chloroform extract were screened against gentamicin induced nephrotoxicity in rats. It was observed that all the test samples except the chloroform extract and a compound (8-dehydroxychrysophanol) isolated and characterised from chloroform extract, decreased the gentamicin induced nephrotoxicity in rats (Aslam, 2011; Aslam, 2012).
7. **Conclusion:** Emodin and 8-dehydroxychrysophanol were characterised in the methanol and chloroform extract of rhizomes respectively. The pure compound, namely 8-dehydroxychrysophanol pesent in chloroform extract may be responsible for enhancing the gentamicin induced nephrotoxicity when treated with chloroform extract in rats.

REFERENCES

Anon. (1972 reprinted 2009). *The Wealth of India: A Dictionary of Indian Raw Materials and Industrial Products. Raw Materials,* Vol. IX:Rh-So, pp. 3–4. National Institute of Science Communication and Information Resources, Council of Scientific and Industrial Research, Publications and Information Directorate, New Delhi, India.

Aslam, M., Dayal, R., Javed, K., Fahamiya., Mujeeb, M., Husain, A. (2012). Pharmacognostical and phytochemical evaluation of *Rheum emodi* Wall. *Current Pharma Research,* 2(2): 471–479.

Aslam, Mohd. (2011). Studies on Some Phytochemical and Biological Medicinal Plants Used in Kidney Disorders, Ph.D thesis, p. 148–160; 266–288. Forest Resarch Institute University (ICFRE), Dehradun, India.

(Y) *Acacia catechu* Willd

1. **Name of the plant species:** *Acacia catechu*
2. **Family:** Mimosaceae
3. **Popular name:** Khair
4. **About the plant:** The genus *Acacia* Mill. includes shrubs, trees and sometimes climbers, spread mostly in the warmer and drier regions of the world, mainly in Africa and Australia. About 40 species are of Indian origin and *Acacia catechu* is one of them. Some more species have been introduced from Africa and Australia depending on their suitability and use. The different *Acacia* species are useful sources of gum, tannins and timber. *Acacia catechu* is a moderate–size deciduous tree possessing a light feathery crown. It occurs throughout India on variety of soils and normally attains the height up to 3 m and girth over 0.9 m. *Acacia catechu* is widespread throughout the sub-Himalayan tract from Punjab to Assam, ascending to an altitude of 1,200 m. It is frequent in the peninsular region, particularly in the drier parts. The *khair* plantations is being raised by direct sowing of seeds (Anon., 1985, reprinted 2010).
5. **Medicinal and other uses**: The heartwood of *Acacia catechu* is of commercial value and is being used for the production of *katha* and *cutch*. The two products are obtained from the heartwood of 20–30 years old trees, and about 0.3 m in girth. The bark and sapwood are used as fuel.

   The constituents of *katha* and *cutch* are catechin and catechutannic acid, respectively. Katha is being used in making of *pan* and medicine in India. It is astringent in nature, cooling and digestive, finds use in relaxed conditions of throat, mouth, and gums, also in cough and diarrhoea. It is also used externally on ulcers, boils and eruption of skin. It is recorded to be antileprotic drug. Cutch is being used in silk, wool, cotton, pulp and paper dyeing. It may also be used for dyeing khaki. Good quality of gum is obtained from *Acacia catechu* which may be used as a substitute for true gum Arabic. The heartwood of *Acacia catechu* constitute a very durable timber and is resistant to white ants and *Teredo* Linn. (Anon., 1985, reprinted 2010).
6. **Reaearch contribution of Sharma, Dayal and Ayyar**:
   - **(a) Numbers of compounds isolated and characterised from leaves, bark and roots:** Sixteen
   - **(b) Names of compounds**

Compounds from Leaves

(i) Poriferasterol
(ii) Quercetin
(iii) Poriferasterol-3-β-D-glucoside
(iv) Quercetin-3-O-arabinofuranoside
(v) Quercetin-3-O-rhamnoside
(vi) Quercetin-3-O-galactoside

Compounds from Bark and Roots

(vii) Lupenone
(viii) Lupeol
(ix) Poriferasterol
(x) Octacosanoic acid
(xi) Poriferasterol acylglucosides (Mixture of four comounds)
(xii) Hexacosanoic acid ester
(xiii) Ursolic acid
(xiv) Poriferasterol-3-β-D-glucoside

Compounds from Roots

(xv) 3-Methylquercetin
(xvi) Quercetin
(xvii) Dihydroquercetin
(xviii) Poriferasterol
(xix) (-)-Epicatechin
(xx) Catechin

(c) **Mode of structure elucidation:** The structures of sixteen compounds as given under 6(b) were elucidated by colour reactions, derivatisation, spectral data(IR, UV, ^{1}HNMR, ^{13}CNMR, Mass), hydrolytic studies and direct comparison with authentic samples wherever available (Sharma et al., 1997; Sharma et al., 1999). The structures of the compounds are given at the end of the chapter.

7. **Conclusion:** The chemical examination of the leaves, bark and roots was undertaken to find their utilisation after the feeling of *Acacia catechu* tree for the heartwood which is used for isolation of *katha* and *cutch*. Sixteen compounds were characterised and a number of these were reported for the first time from this *Acacia* species.

REFERENCES

Anon. (1985 reprinted 2010). *The Wealth of India: A Dictionary of Indian Raw Materials and Industrial Products. Raw Materials*, Vol. I:A, pp. 23–31. National Institute of Science Communication and Information Resources, Council of Scientific and Industrial Research, Publications and Information Directorate, New Delhi, India.

Sharma, P., Dayal, R., Ayyar, K.S. (1997). Chemical constituents of *Acacia catechu* leaves. *J. Indian Chem. Soc.*, 74(1): 60.

Sharma, P., Dayal, R., Ayyar, K. S. (1999). Acylglucosterols from *Acacia catechu* (roots), *J. Medicinal and Aromatic Plants*. 21(4): 1002–1005.

(Z) *Dicentra paucinervia* Stern

1. **Name of the plant species:** *Dicentra paucinervia*
2. **Family:** Fumariaceae
3. **Popular name:** The common name, Bleeding heart, is used for many *Dicentra* species.
4. **About the plant:** *Dicentra* Bernh is a genus of slender herbaceous flowering plants with tuberous root stocks, scattered in temperate northern Asia and northern America (Anon, 1952, reprinted 2010). The genus is reported to have twenty species (Aggarwal et al., 2009). Six *Dicentra species,* namely, *D. canadensis, D.chrysantha, D. formosa, D. spectabilis, D. paucinervia* and *D. macrocapnos* have been reported in India (Gaur, 1999). A number of physiologically active isoquinoline alkaloids (aporphines, protopberberines, protopines and cularines alkaloids) have been identified in the tubers of different *Dicentr*a species.

   Dicentra paucinervia, a highly potent ethnomedicinal herb, is distributed in Arunachal Pradesh, Meghalaya, Nagaland, Sikkim and West Bengal in India. It exhibits dual habits, i.e., a climber when supports are given or also creeper. Its leaves are repeatedly ternate with terminal leaflets usually replaced by a branched tendril. The plants usually grow well in loose moist sandy loam or loamy or forest soil rich in organic matters up to length of 6 feet. The tubers are collected in the month of December (Aggarwal, 2009a; Aggarwal et al., 2009).
5. **Medicinal and other uses**: The plant is of medicinal value and has been used by Naga ethnic tribal communities in eastern Nagaland and adjoining Manipur state mainly in the treatement against various diseases like diarrhoea, diabetes, dysentery, malaria, pneumonia, stomach disorders, typhoid and other common fevers, carminatives/flatulence, and cut/injury (Aggarwal et al., 2009).
6. **Research contribution of Aggarwal, Varshney, Dayal and Puni:**
   (a) **Numbers of compounds isolated and characterised from tubers and leaves:** Two

(b) Names of compounds:

(i) Protopine

(ii) Allocryptopine

(c) Mode of structure elucidation: The structures of pure isolated compounds *viz.*, protopine and allocryptopine were elucidated using IR, UV, ^{1}HNMR, ^{13}CNMR, and Mass spectra. A sensitive, simple, rapid and accurate High Performance Thin Layer Chromatography (HPTLC) method was developed for the quantitation of protopine and allocryptopine in tubers (Aggarwal et al., 2009; Aggarwal, 2009a). The amount of protopine was estimated to be 0.51 per cent and 0.67 percent from the tubers grown in Manipur and cultivated in Dehradun, respectively. The structures of the compounds are given at the end of the chapter.

(d) Pharmacological and other activities: Protopine and allocryptopine were screened for their anti-allergic, anti-obesity, anti-inflammotry and anti-bacterial activites. No activity was observed for both the compounds (Aggarwal, 2009a).

7. **Conclusion:** Protopine and allocryptopine were characterised for the first time from the medicinally important *Dicentra paucinervia* tubers and leaves. A method was standardised for the estimation of protopine and allocryptopine in the tubers by High Performance Thin Layer Chromatography (HPTLC).

REFERENCES

Anon. (1952 reprinted 2010). *The Wealth of India: A Dictionary of Indian Raw Materials and Industrial Products. Raw Materials*, Vol. III:D-E, p. 53. National Institute of Science Communication and Information Resources, Council of Scientific and Industrial Research, Publications and Information Directorate, New Delhi, India.

Aggarwal, P., Varshney, V.K., Dayal, R. and Puni, L. (2009). Isoquinoline alkaloids from ethnobotanically important *Dicentra paucinervia*. *Int. J. Chem.Sci*; 7(3):1841–1850.

Aggarwal (2009a). Phytochemical examination of *Dicentra paucinervia*, Ph.D. thesis, Forest Resarch Institute University (ICFRE), Dehradun, India.

Gaur, R.D. (1999). Flora of the district–Garhwal North West Himalaya (with ethnobotanical notes) Srinagar (Garhwal), pp. 82–83.

(Z_1) *Asparagus racemosus* Willd

1. **Name of the plant:** *Asparagus racemosus*
2. **Family:** Asparagaceae, Liliaceae

3. **Popular name**: *Satavari, Satawar*
4. **About the plant:** *Asparagus* is a genus of under-shurbs or herbs scattered in the temperate and tropica section of the Old world. The share of India is about 22 species. The roots of some of the species are the sources of useful drug named as *Shatawar. Asparagus racemosus* occurs as wild in tropical areas of India including the Andamans; and climbing in the Himalayas up to an altitude of 1500 m. It is a spinous, highly branched perennial climber, with tuberous, short rootstock and whitish grey or brown woody stems (Anon., 1985, reprinted 2010). The black cotton soil mixed with river sand is suitable for the propagation of the plants. It occurs commonly in the upper Gangetic plains and the Bihar Plateau. The plants bear the white fragrant flowers after the rains. The roots of the plants are of medicinal value and sold in pieces 5–15 cm in length and 2 cm in thickness (Anon., 1985, reprinted 2010).
5. **Medicinal and other uses**: The roots of "Satavari" are of medicinal value and are widely used in traditional systems of medicines, such as Ayurveda, Siddha and Unani (Charak Samhita, 1970). The roots find use in the treatment of several ailments including dysentery, tumours, inflammations, neuropathy, certain infectious diseases, epilepsy, fatigue and tuberculosis (Sharma et al., 2000). The other medicinal uses have also been reported (Chamoli et al., 2013). A number of pure chemical compounds have also been isolated and characterised from the roots including the steroidal saponins (Shatavarin 1-X) of which Shatavarin IV, a major compound, is the marker compound of the drug.
6. **Research contribution of Chamoli et al:**
   (a) **Number of compounds isolated and characterised from the roots:** One
   (b) **Name of compound:** Shatavarin IV
   (c) **Mode of structure elucidation:** Shatavarin IV (Sarsasapogenin glycosides having two molecules of rhamnose and one molecule of glucose) was isolated from the n-butanol extract of the roots by column chromatography and its identity was confirmed by direct comparison with authentic sample (Co-TLC, mmp and superimposable IR spectra). The Shatavarin IV content in the roots of the samples of *A. racemosus* grown as intercrop and solo crop was estimated by a self developed standardised High Performance Thin Layer Chromatography (HPTLC) method (Chamoli et al., 2013). The structure of shatavarin IV is given at the end of the chapter.
7. **Conclusion:** A simple, rapid, specefic, and precise HPTLC method was developed and validated for the quantitative estimation of shatavarin IV in the roots of *A. racemosus* grown as intercrop (0.38%), in mulberry plantations and sole crop (0.39%).

REFERENCES

Anon. (1985, reprinted 2010). The wealth of India: *A Dictionary of Indian Raw Materials and Industrial Products. Raw Materials,* Vol. I:A, p. 469–471. National Institute of Science Communication and Information Resources, Council of Scientific and Industrial Research, Publications and Information Directorate, New Delhi, India.

Chamoli, M., Varshney, V.K., Srivastava, P.K., Pandey, R., Dayal, R. (2013). HPTLC Method Development and Analysis of Bioactive marker Shatavarin IV in *Asparagus racemosus* Intercropped with *Morus alba. TACL,* 13(2): 125–138.

Charak Samhita. (1970). Volume 1–V, *Chowkhamba Sanskrit Series,* Orientalia, Varnasi, India.

Sharma, P.C., Yelene, M.B. and Dannis, T.J. (2000). *Database on Medicinal Plants Used in Ayurveda,* Vol. 1, CCRAS, New Delhi, p. 418.

(Z_2) *Acorus calamus* Linn

1. **Name of the plant:** *Acorus calamus*
2. **Family:** Araceae
3. **Popular name:** Bach, Sweet Root, Sweet Flag
4. **About the plant:** *Acorus* is a genus of herbs, raised in the North temperate and sub-tropical areas. Two species, namely, *Acorus calamus, Acorus gramineus,* of medicinal valuc have been recorded in India. *Acorus calamus* is a semi-aquatic, perennial, aromatic herbaceous plant having creeping rhizomes. It occurs wild and also being cultivated throughout India, soaring to an altitude of 2,200 m in the Himalayas. Sweet root grows best in marshy and moist areas. It abundantly occurs in the marshy tracts of Srimaur (Himachal Pradesh), Kashmir, Manipur and the Naga hills. The clayey loams and light alluvial soils are normally suitable for its cultivation (Anon., 1952, reprinted 2010).
5. **Medicinal and other uses:** The dried rhizomes are sold as drug in the market. The rhizomes are believed to have anti-spasmodic, carminative, and anthelmintic properties, and are used for the treatment of a number of diseases like epilepsy, and other mental ailments, chronic diarrhoea and dysentery, bronchial catarrh, intermittent fevers and glandular and abdominal tumours. The rhizomes are also used for liver and kidney troubles, eczema and rheumatism (Anon., 1952, reprinted 2010).
6. **Research contribution of Chamoli et al.**:
   (a) **Number of compounds characterised in the essential oil from rhizomes:** One
   (b) **Name of compound:** β-Asarone

(c) **Mode of structure elucidation:** The essential oils yield in the rhizomes of *Acorus calamus* intercropped with *Morus alba* and of control samples were determined by hydrodistillation method using Clevenger-type apparatus. (Chamoli et al., 2013). The identification of *β*-asarone in the oil was done by GLC using a Chemito GC1000 gas chromatographed equipped with a BP 5 capillary column (30 m × 0.22 mm; film thickness 0.25 μm) and Flame Ionisation Detector. The other conditions of the analysis of the oils have also been described. The essential oil yield was found to vary from 1.55 per cent (control) and to 1.57 per cent (intercropped), while their *β*-asarone content varied from 87.32 per cent (control), 93.03 per cent (intercropped). The structure of asarone is given at the end of the chapter.

7. **Conclusion:** The statistical analysis of the data indicated, no vital harmful allelopathic effect of *Morus alba* on the biomass yield, essential oil and *β*-asarone content of *Acorus calamus* intercropped with *Morus alba*.

REFERENCES

Anon. (1985, reprinted 2010). *The Wealth of India: A Dictionary of Indian Raw Materials and Industrial Products. Raw Materials*, Vol. I:A, pp. 63–65. National Institute of Science Communication and Information Resources, Council of Scientific and Industrial Research, Publications and Information Directorate, New Delhi, India.

Chamoli, M., Varshney, V.K., Srivastava, P.K., Pandey, R., Dayal, R. (2013). Assesment of Biomass Yield, Essential oil and *β*-asarone content *of Acorus Calamus L.* intercroped with *Morus alba L. J.Essential Oil Bearing Plants,* 16(6): 763–770.

(Z_3) *Cryptomeria japonica* (Linn. f.) D. Don

1. **Name plant species**: *Cryptomeria japonica*
2. **Family:** Cupressaceae, Taxodiaceae
3. **Popular name:** Japanese cedar, Japan cypress
4. **About the plant:** *Cryptomeria japonica,* an exotic plant species from Japan, is an evergreen tree cultivated largely in the forestry plantations in the Darjeeling district and Sikkim in India, and known as *dhuppi*. It is regularly found in Japan, where it is commonly known as *sugi*. It is also found in China and *Azores* islands. It grows up to 70 m in height and 4.0 m trunk diameter, having red–brown bark which can be peeled in vertical strips (http://en.Wikipedia). The Indian trees grow faster than the Japanese trees with the average annual increments of 20 to 21 m^3 per hectare per annum (Anon, 2001 reprinted 2010), while Japanese wood has 10 to 12 rings per inch diameter, the Indian wood has only two rings or evenless (Anon., 2001, reprinted 2010; Gupta et al., 1987).

5. **Medicinal and other uses**: *Cryptomeria japonica* yields low density, fragrant, weather and insect resistant wood which may be used for making boxes, veneers, plywood, furniture and for other purposes. It is largely used for house building in Sikkim and Darjeeling district of India (https://en.Wikipedia). The essential oil and other heartwood constituents of the Japanese tree have been extensively studied (Gupta et al., 1987), but the essential oil constituents of Indian trees remained unexplored till 1987.

6. **Research contribution of Gupta, Dayal and Ayyar**:

   (a) **Number of compounds characterised in the essential oil of heartwood:** Eight

   (b) **Names of compounds characterised:**

   (i) α-Calacorene
   (ii) α-Muurolene
   (iii) Calamene
   (iv) β-Calacorene
   (v) Epicubenol ester
   (vi) Torreyol
   (vii) Epicubenol
   (viii) Cubenol

   (c) **Mode of structure elucidation:** The acetone extract of the powdered heartwood on steam-distillation gave an essential oil (1%). The structures of the above cited eight compounds from the essential oil were elucidated by using gas liquid chromatography (GLC), isolating them pure by column chromatography wherever possible, followed by their structure elucidation using spectral data (IR, UV, ^{1}HNMR and Mass)[Gupta et al., 1987]. The structures of compounds are given at the end of the chapter.

7. **Conclusion:** The studies indicated that the three identified compounds namely (-)-epicubenol, (-)-cubenol, and(-)-epicubenol ester taken together constituted about 32 per cent of the neutral fraction of the essential oil of Indian *C.japonica* and have not been reported to occur in the essential oil of Japanese trees. Also, more than 94 per cent of the Indian oil consists of sesquiterpenes with cadalene skelton. On the other hand, the Japanese essential oil contains cryptomerione, a bisabolene sesquiterpenoid ketone and β–eudesmol having eudesmane skelton. The investigation indicated a difference in composition of the essential oil of the heartwood of *C. japonica* growing in Japan and India.

REFERENCES

Anon. (2001 reprinted 2010). The wealth of India: *A Dictionary of Indian Raw Materials and Industrial Products. Raw Materials*, Vol. II:Cl-Cy, First Supplement Series, p. 251. National Institute of Science Communication and Information Resources, Council of Scientific and Industrial Research, Publications and Information Directorate, New Delhi, India.

Gupta, K., Ayyar, K. S., Dayal, R. (1987). Essential Oil of Indian *Cryptomeria japonica. J. Indian Chem. Soc.*, 64: 66.

https:/en.Wikipedia.org/wiki/Cryptomeria/October 2015. 22.12.2015

Structures of chemical constituents isolated and characterised from twenty nine plant species given in the above order

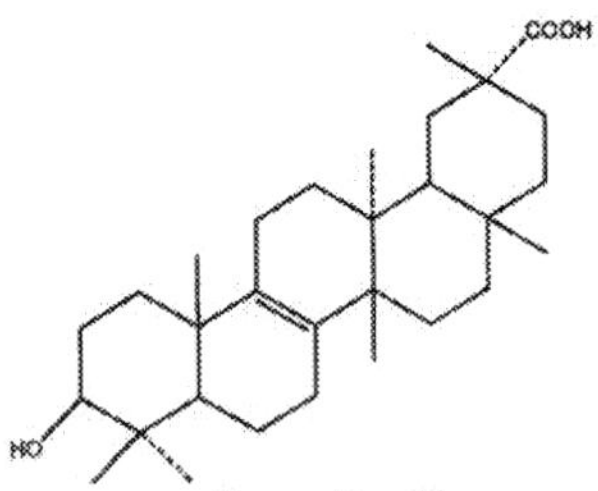

Bryonolic acid

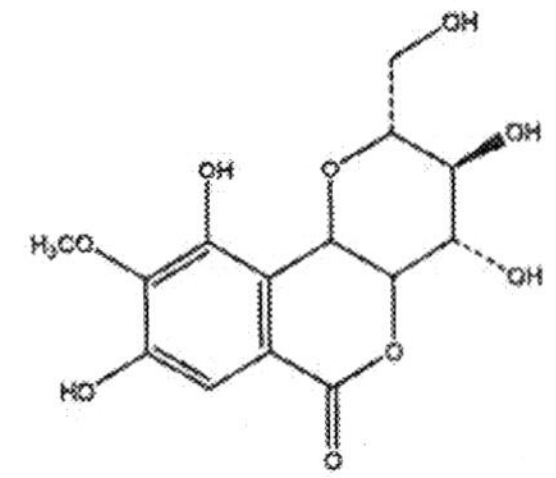

Bergenin

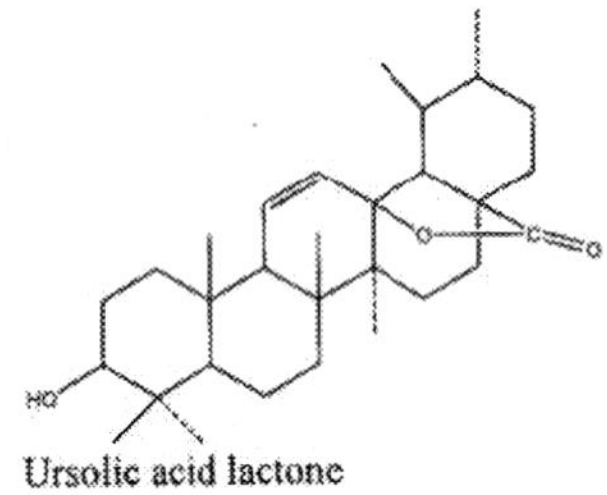

Ursolic acid lactone

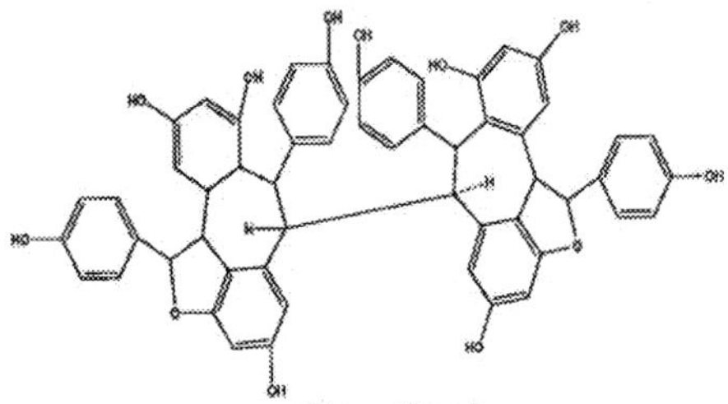

Hopeaphenol

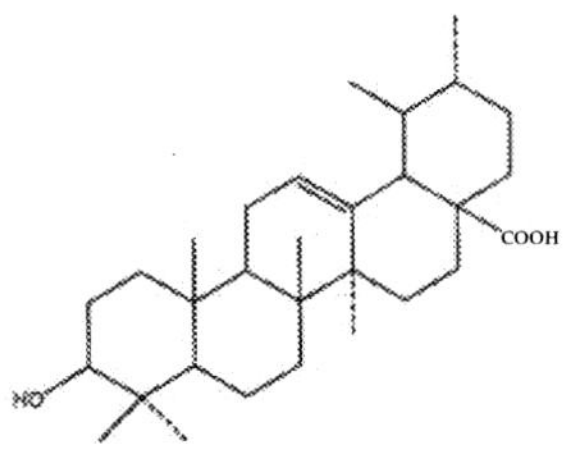

Ursolic acid

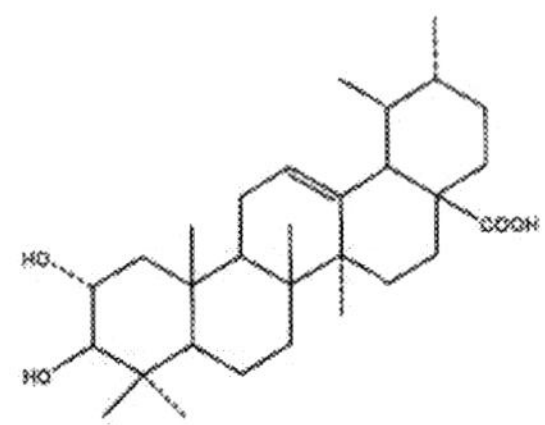

2α-hydroxyursolic acid

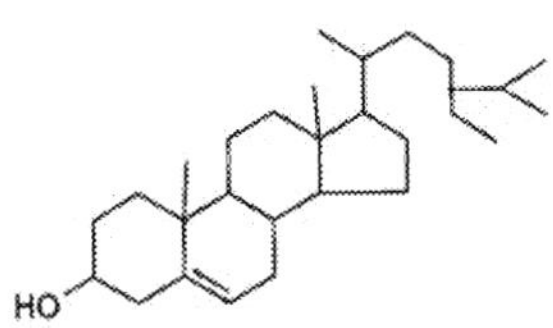

β-Sitosterol

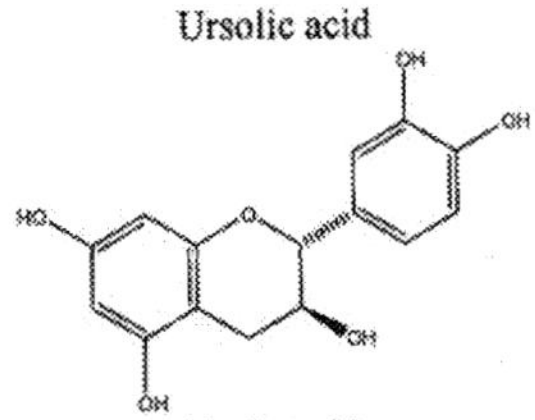

(+)-Catechin

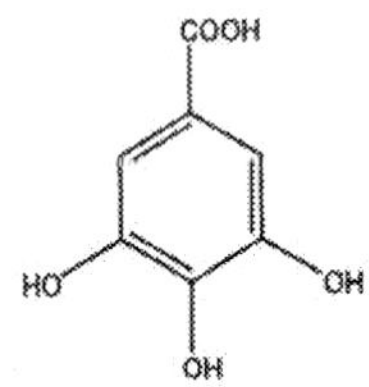

Gallic acid

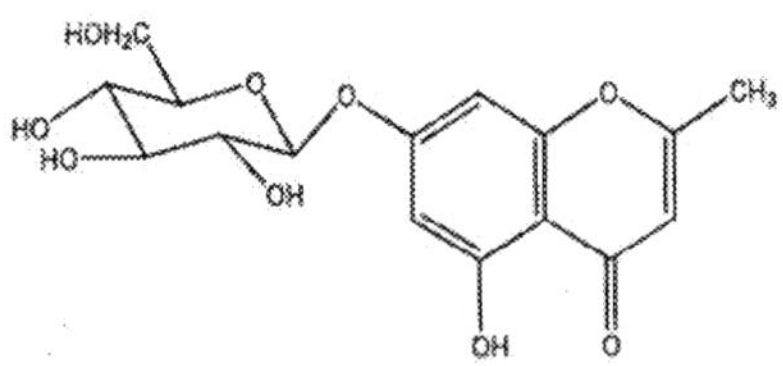

7-β-D-glucosyloxy-5-hydroxy-2-methylchromone

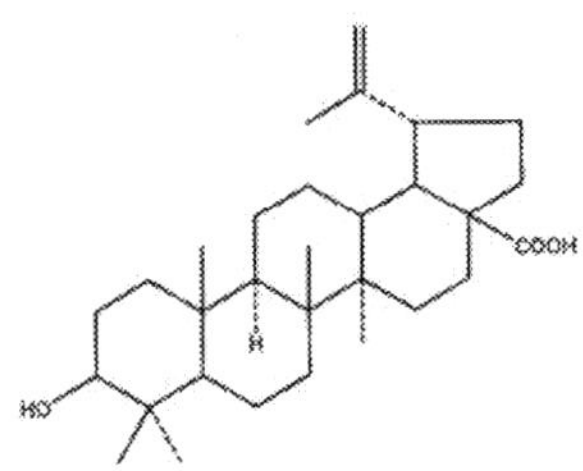

Betulinic acid

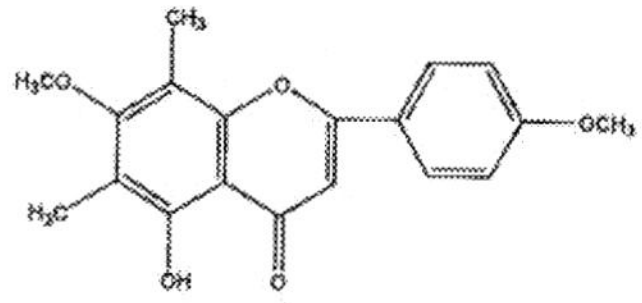

Eucalyptin

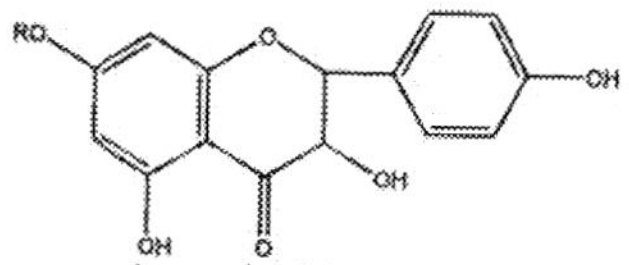

Aromadendrin R=H
7-O-Methylaromadedrin R=CH_3

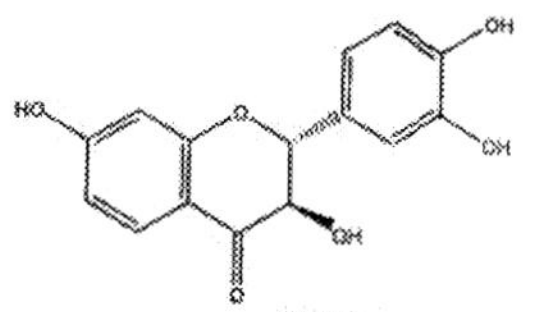

Fustin

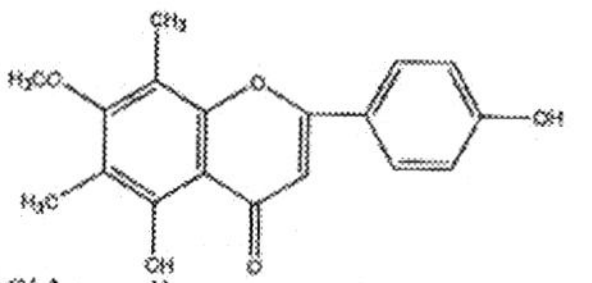

Sideroxylin

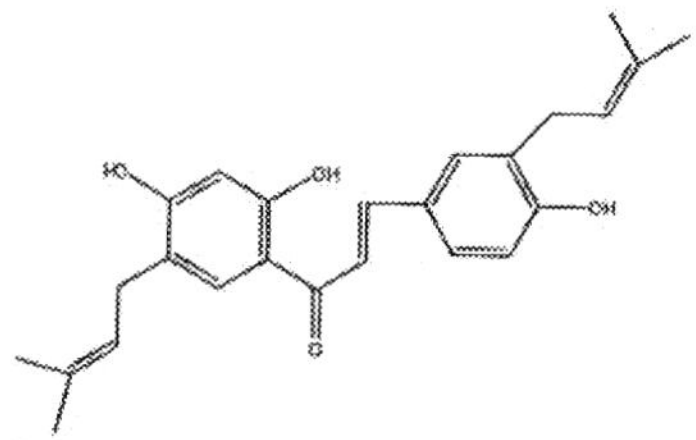

Stipulin

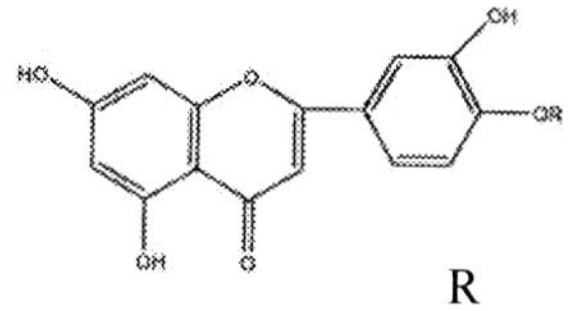

	R
Luteolin	H
Luteolin 4′-O-rutinoside	β-D-glucopyranosyl-(6→1)-α-L-rhamnopyranoside (Rutinosyl)

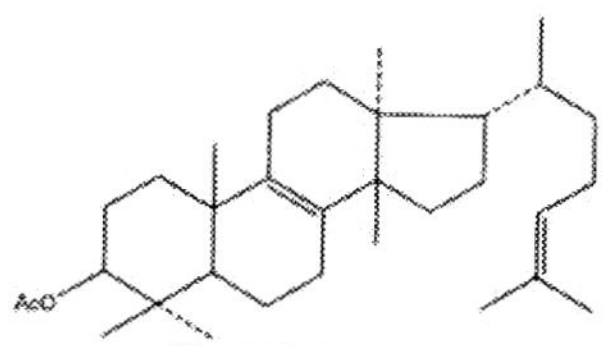

Tirucallol acetate

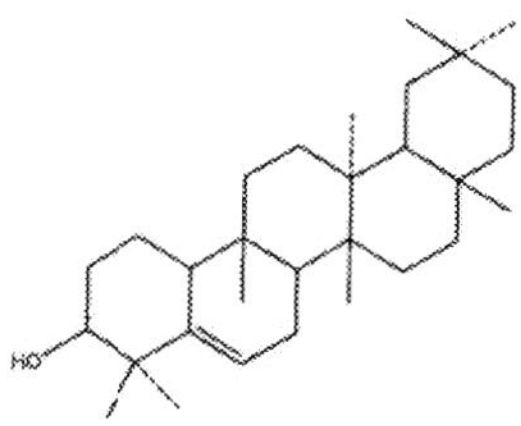

Glutinol

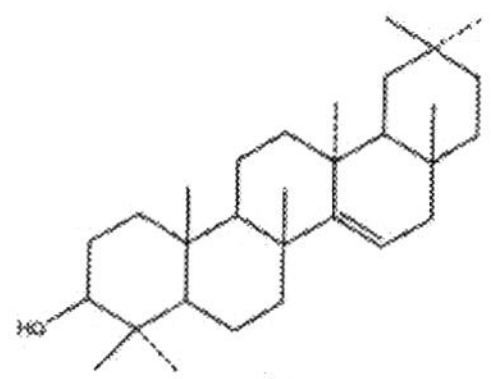

Taraxerol

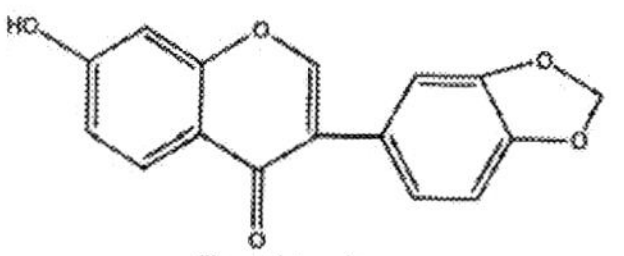

ψ-Baptigenin

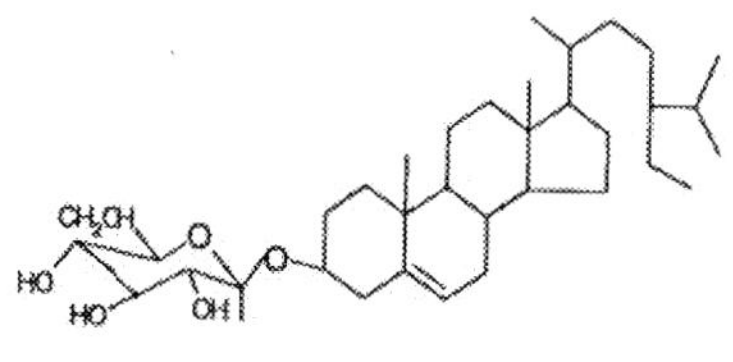

β-Sitosterol-3-O-glucoside

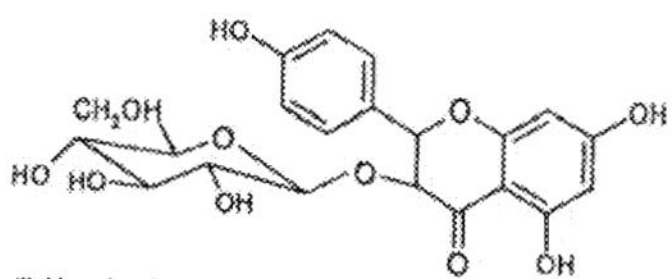

Dihydrokaempferol-3-O-glucoside

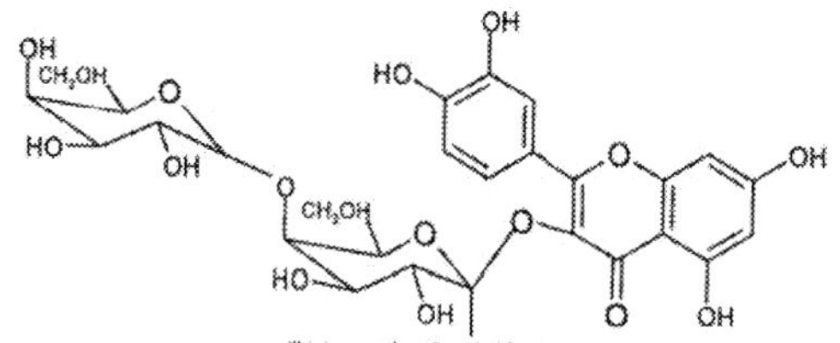

Quercetin-3-O-dialactoside

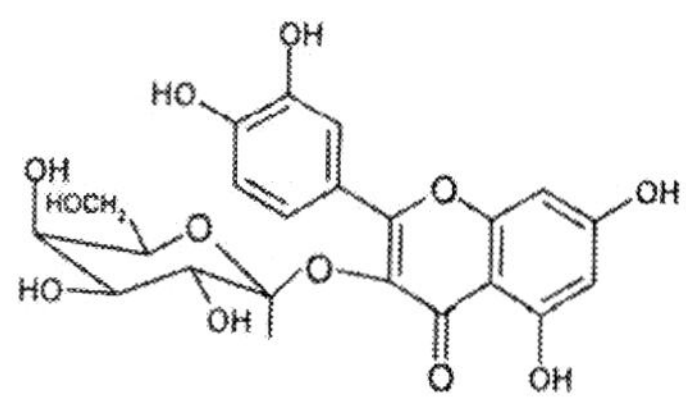

Quercetin-3-O-galactoside

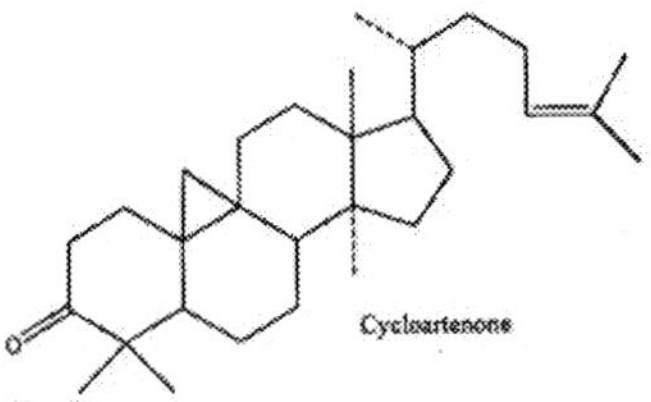

Cycloartenone

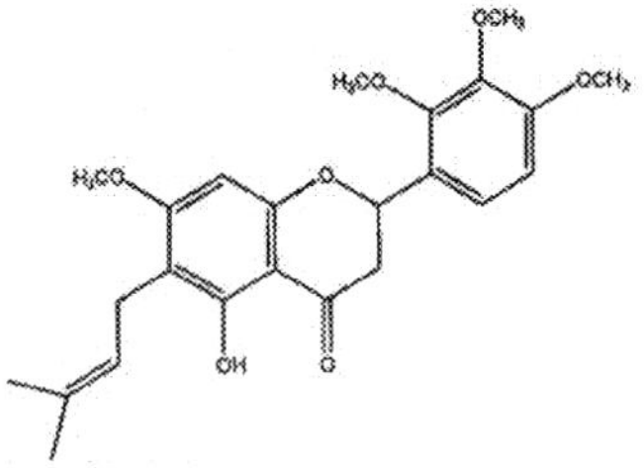

Artoflavanone

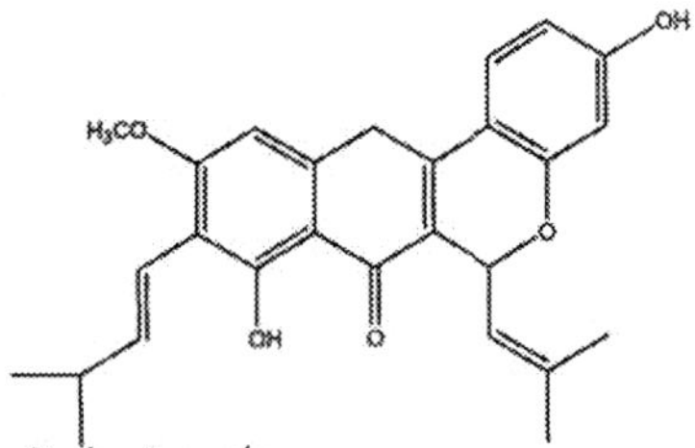

Cycloartocarpin

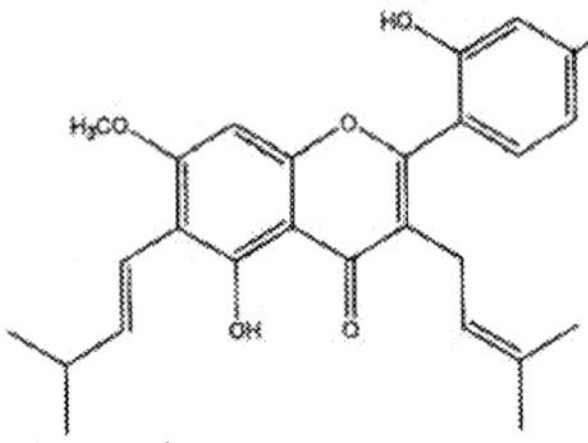

Artocarpin

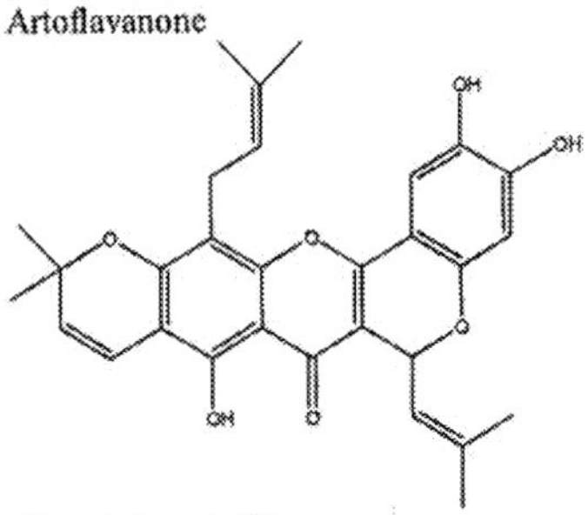

Cycloheterophyllin

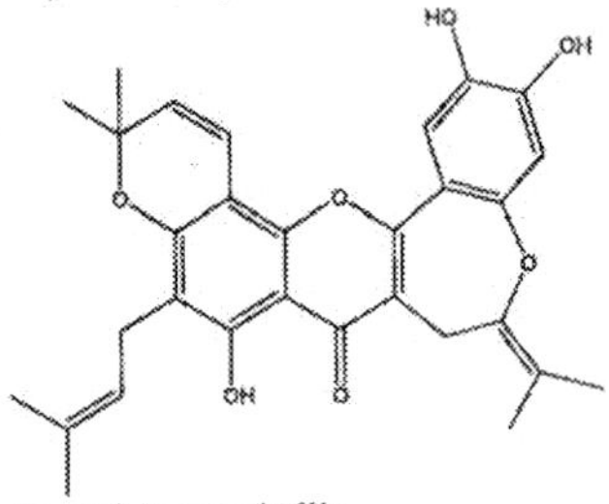

Isocycloheterophyllin

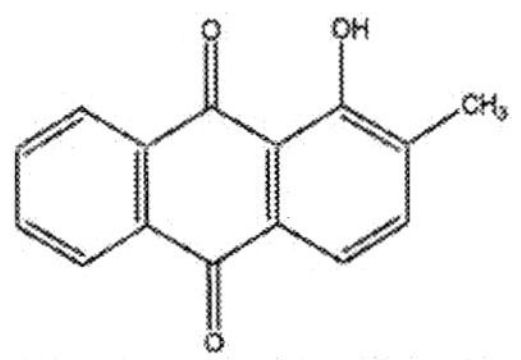

1-hydroxy-2-methylanthraquinone

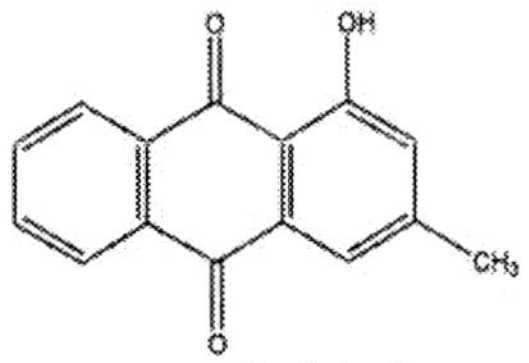

Pachybasin

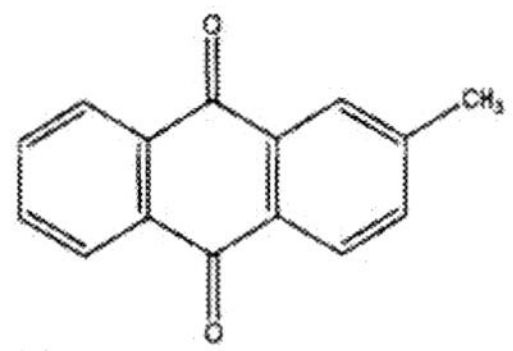

Tectoquinone

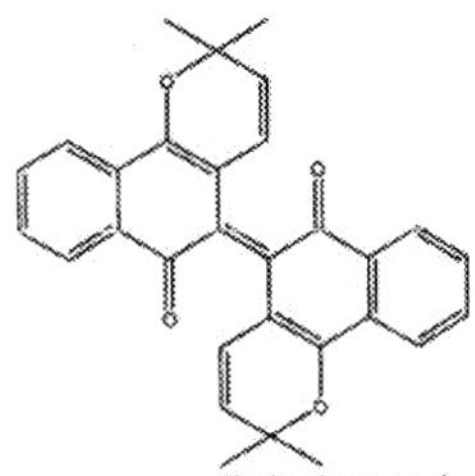

Dehydrotectol

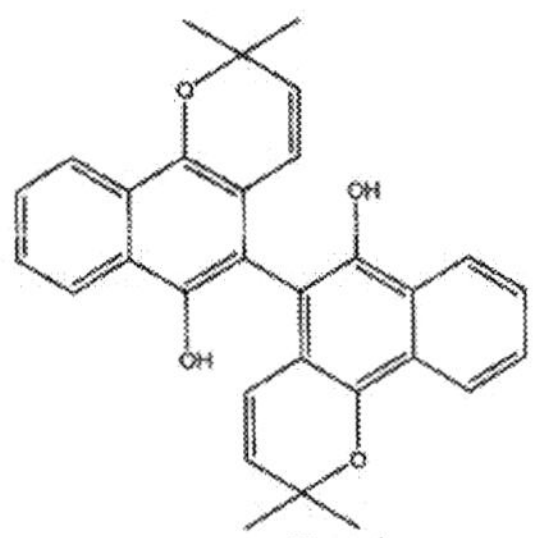

Tectol

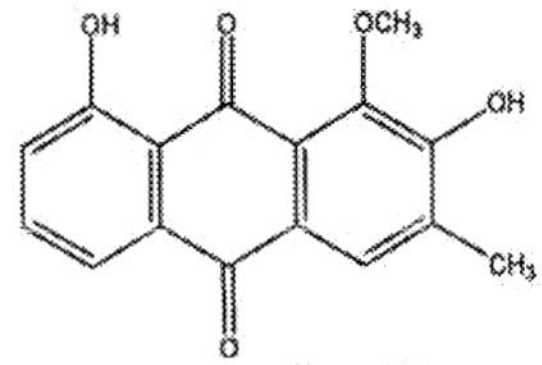

Obtusifolin

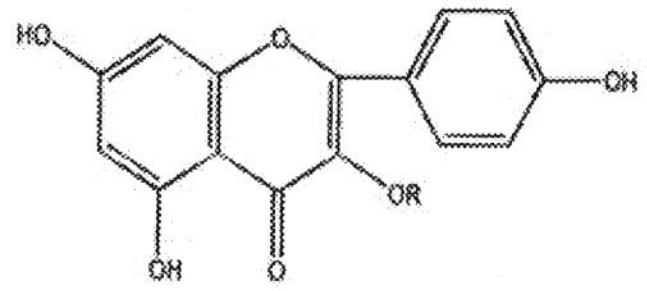

	R
Astragalin	β-D-glucopyranosyl
Trifolin	β-D-galactopyranosyl

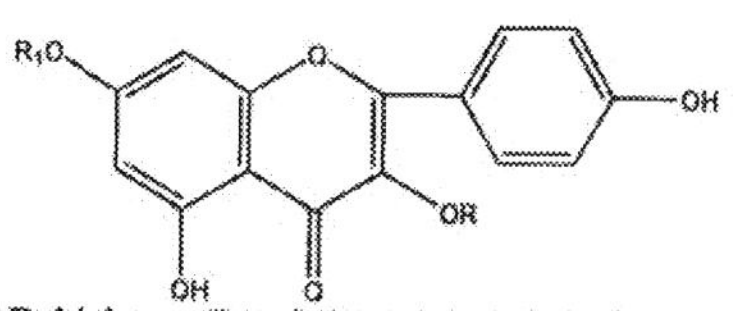

Robinin R= β-D-galactopyranosyl (6→1) O-α- L-rhamnopyranoside
R_1= α-L-rhamnopyranosyl

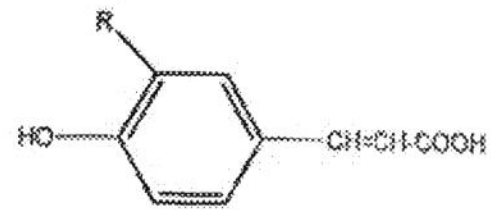

	R
p-Coumaric acid	H
Caffeic acid	OH
Ferulic acid	OCH_3

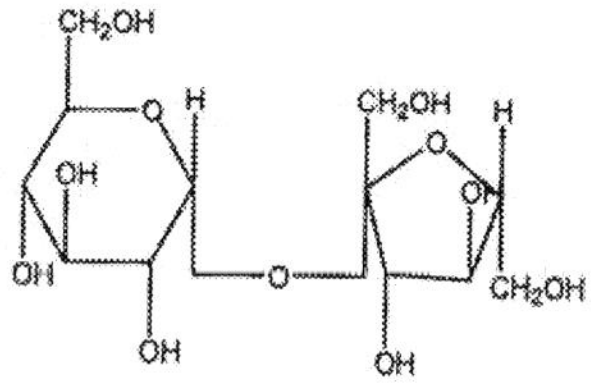

Sucrose

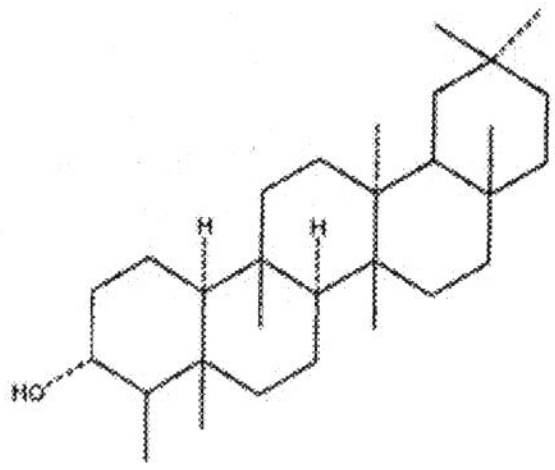

Friedelan-3α-ol

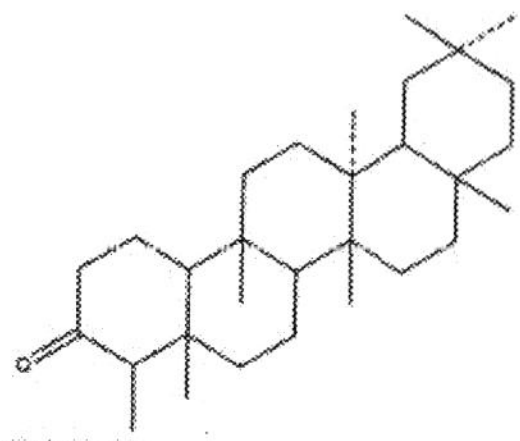

Friedelin

$HO\text{-}CH_2\text{-}(CH_2)_9\text{-}CH_2\text{-}COOH$

12-Hydroxydodecanoic acid

$HO\text{-}CH_2\text{-}(CH_2)_{11}\text{-}CH_2\text{-}COOH$

14-Hydroxy tetradecanoic acid

$HO\text{-}CH_2\text{-}(CH_2)_{13}\text{-}CH_2\text{-}COOH$

16-Hydroxyhexadecanoic acid

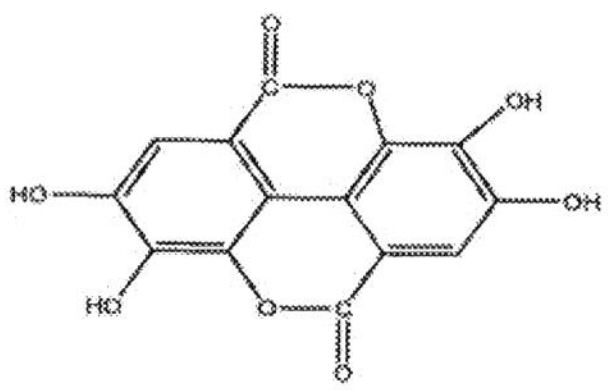

Ellagic acid

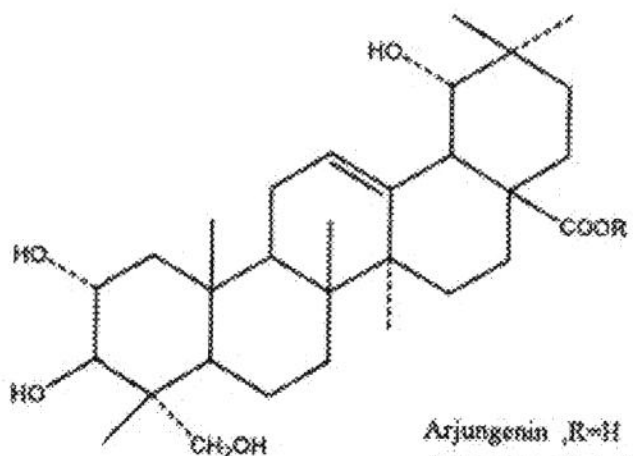

Arjungenin ,R=H
Arjunglucoside , R=Glucosyl

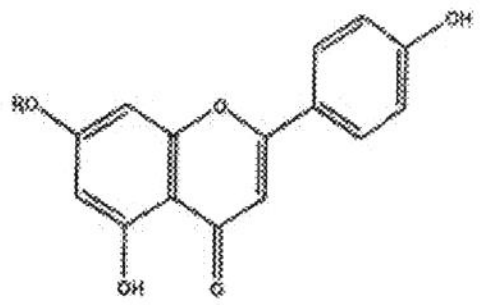

	R
Apigenin	H
Apigenin 7-O-glucoside	Glucosyl

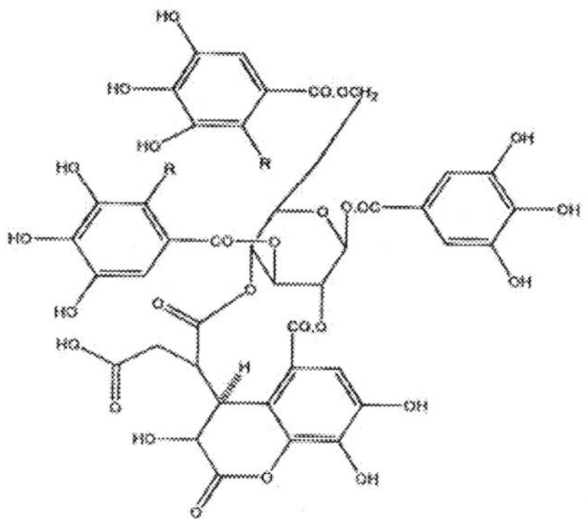

Chebulinic acid
R=H
Chebulic acid
RR=Bond

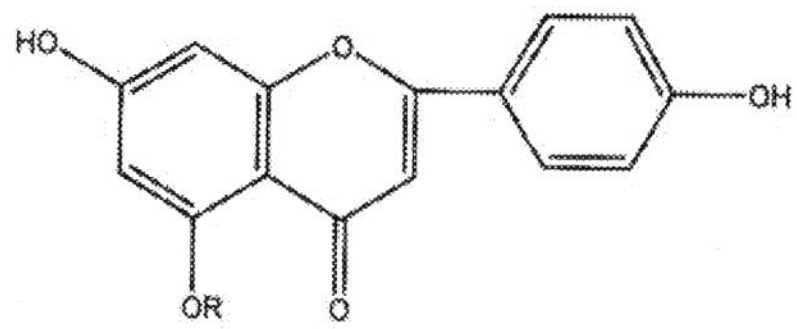

Compound 2: Apigenin 5-O-D-β-glucopyranosyl-(2→1)-α-L-rhamnopyranoside (R= β-D-glucopyranosyl-(2→1)-α-L-rhamnopyranoside)

Compound 3: Apigenin 5-O-β-D-(6″acetylglucopyranosyl-(2→1)-α-L-rhamnopyranoside (R= β-D-(6″acetylglucopyranosyl-(2→1)-α-L-rhamnopyranoside)

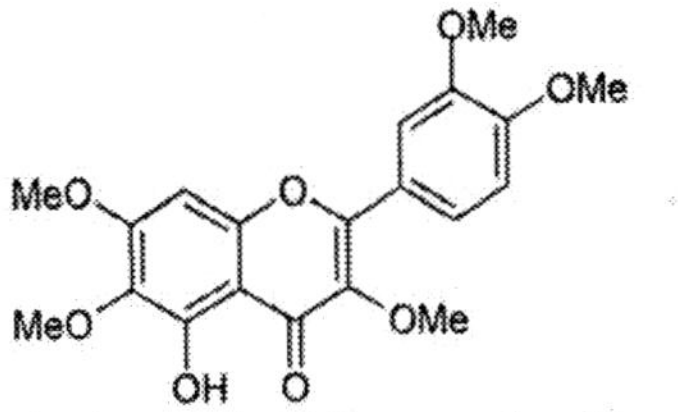

5-hydroxy-3,6,7,3′,4′-pentamethoxyflavone

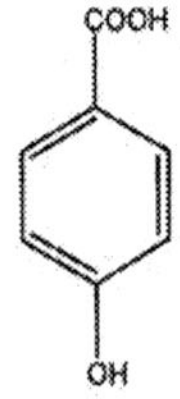

p-Hydroxybenzoic acid

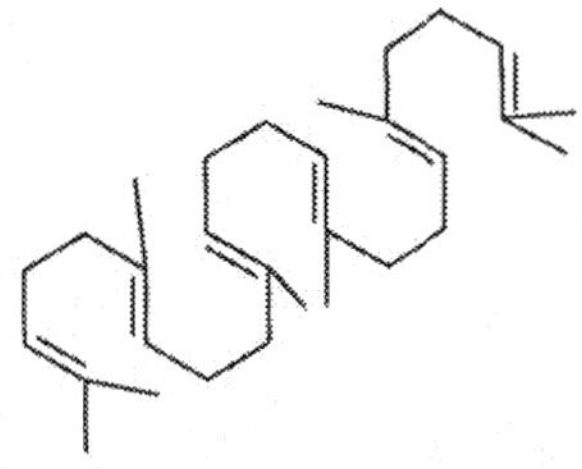

Squalene

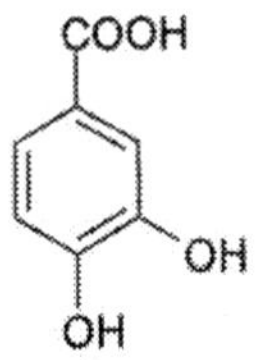

3,4-Dihydroxybenzoic acid

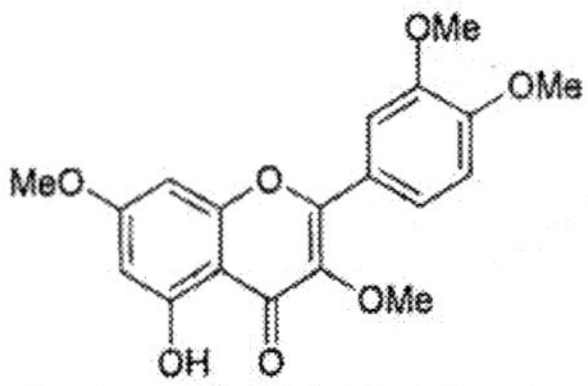

5-hydroxy-3,7,3′,4′-tetramethoxyflavone

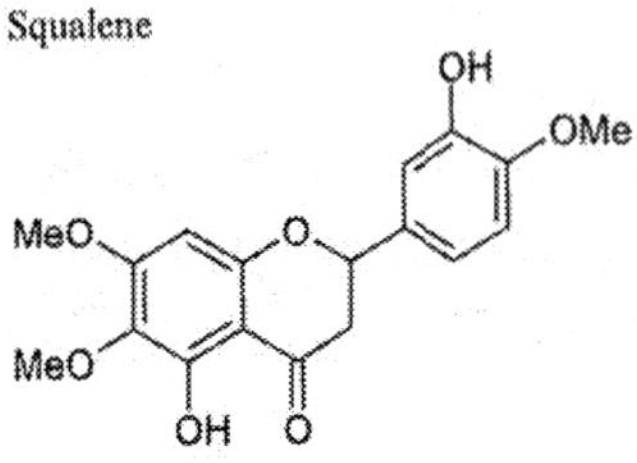

5,3′-dihydroxy-6,7,4′-trimethoxyflavanone

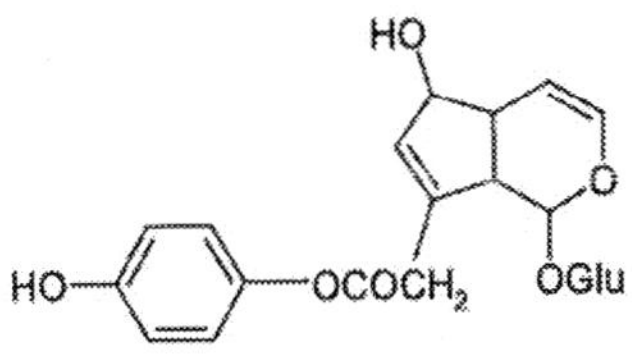

Agnuside

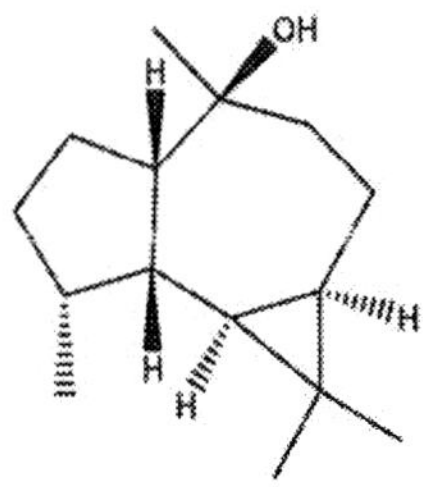

Viridiflorol

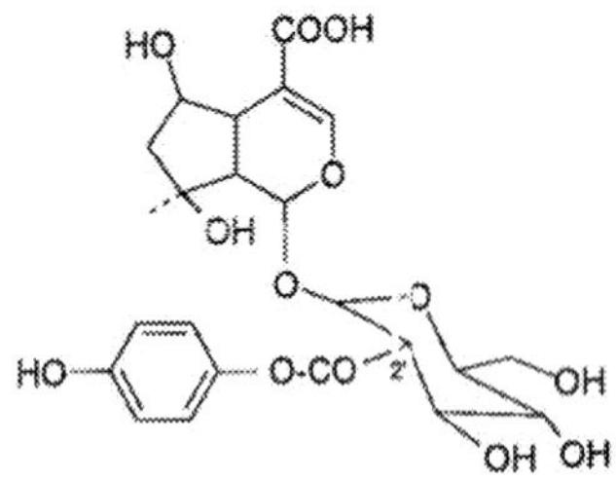

2′-p-hydroxybenzoylmussaenosidic acid

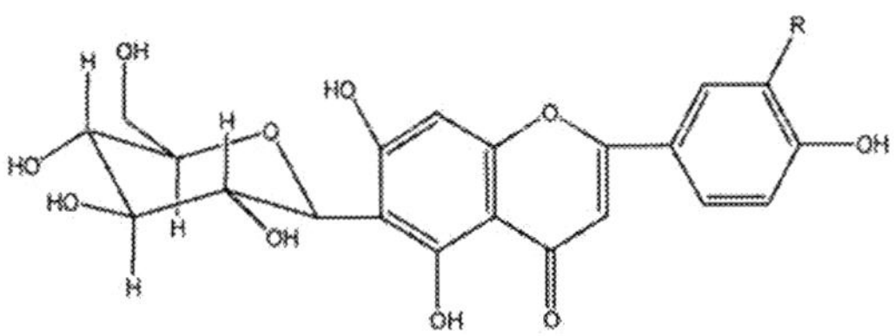

	R
Isoorientin(Homoorientin)	OH
Isovitexin	H

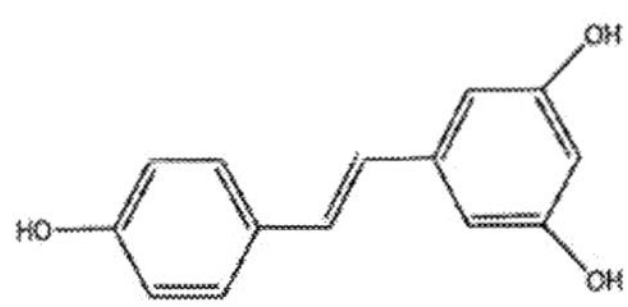

Resveratrol

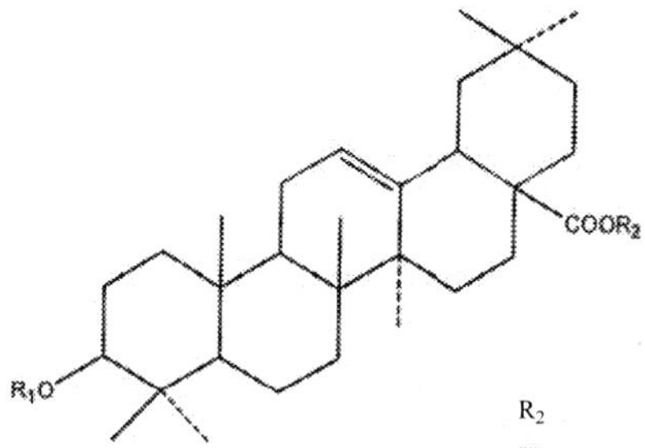

	R$_1$	R$_2$
1	GlcA(4 → 1) Rha	H
2	GlcA(4 → 1) Rha	Glc
3	GlcA(4 → 1) Rha	Glc(4 → 1) Glc

1= Alpha-L-rhamnopyranosyl-(1→4)-(beta-D-glucopyranosyluronic acid)-(1→3)-oleanolic acid

2= Alpha-L-rhamnopyranosyl-(1→4)-(beta-D-glucopyranosyluronic acid)-(1→3)-oleanolic acid-28-O-beta-D-glucopyranoside

3= Alpha-L-rhamnopyranosy-(1→4)-(beta-D-glucopyranosyluronic acid-(1→3)-oleanolic acid-28-O-beta-D-glucopyranosyl-(1→4)-beta-D-glucopyranoside

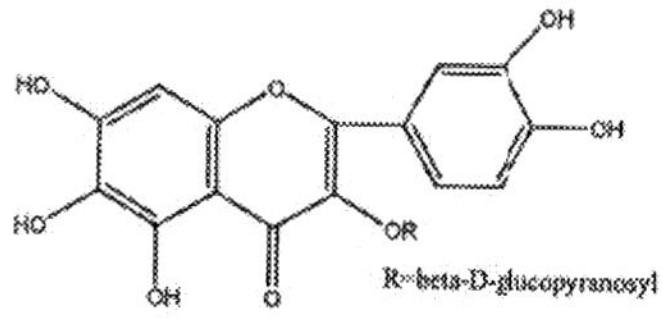

Quercetagetin 3-O-β-glucoside

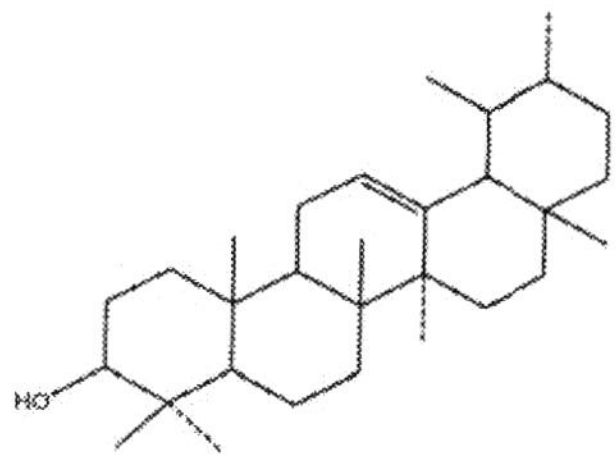

α-Amyrin

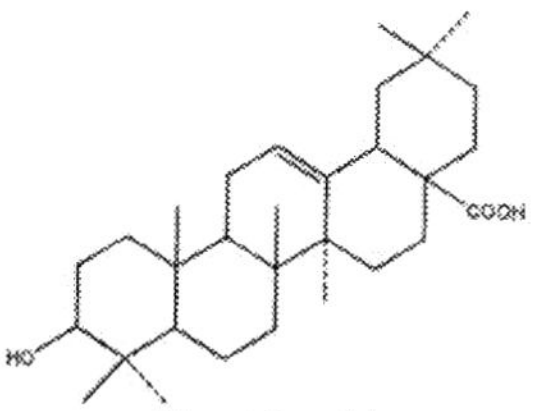

Oleanolic acid

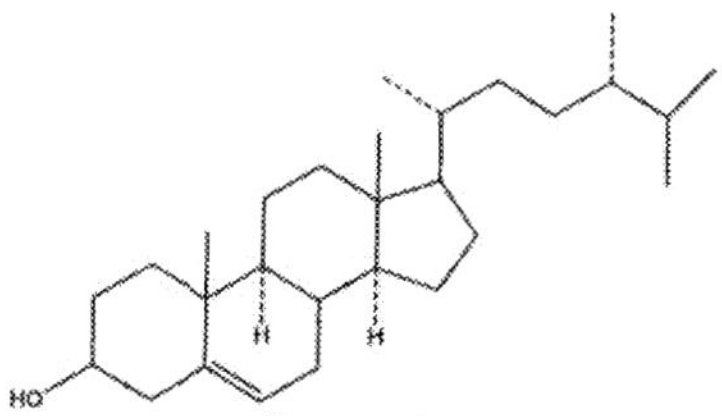

Campesterol

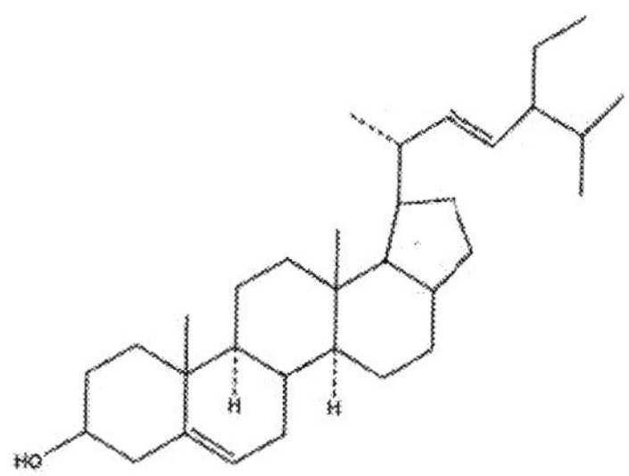

Stigmasterol

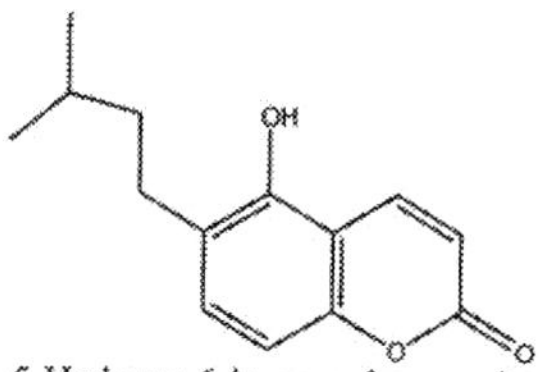

5-Hydroxy-6-isopranylcoumarin

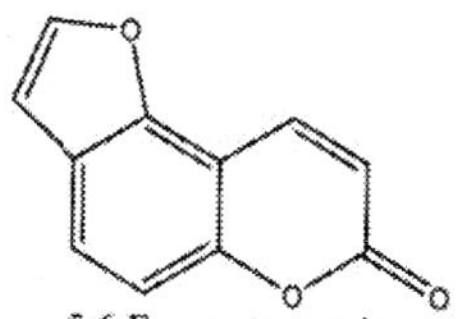

5,6-Furanocoumarin

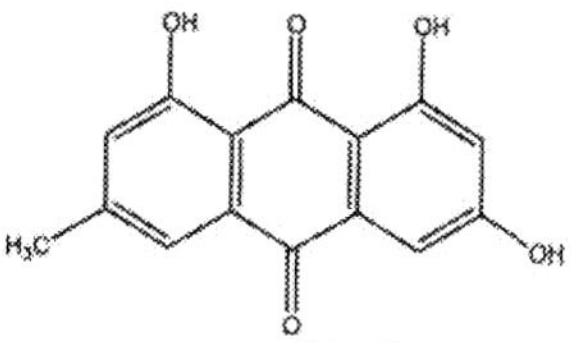

Emodin

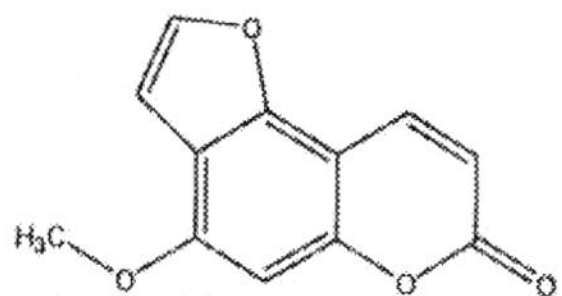

7-Methoxy-5,6-furanocoumarin

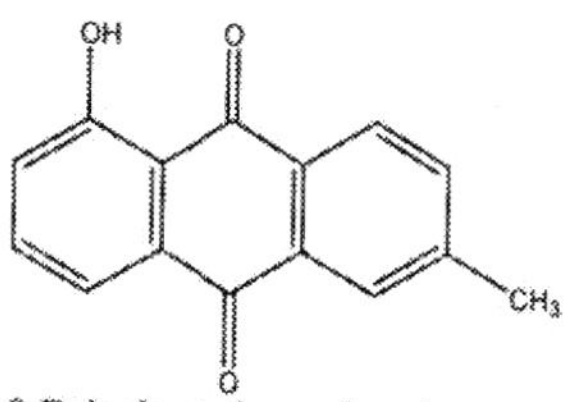

8-Dehydroxychrysophanol

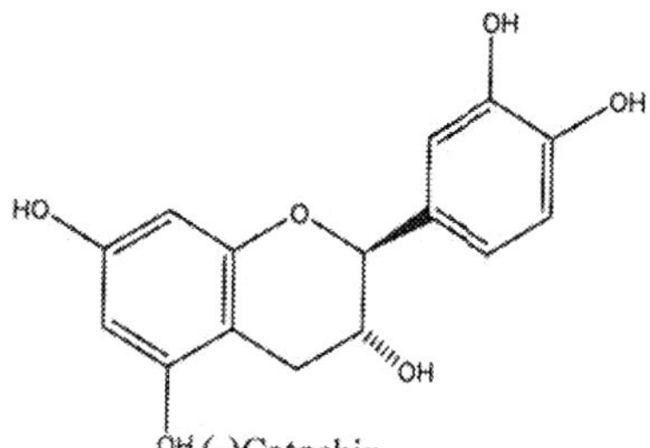

(-)Catechin

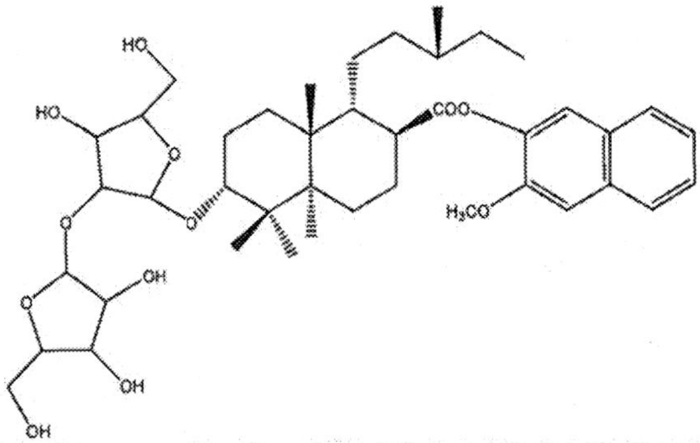

Labdanyl-3α-ol-18-(3‴-methoxy-2‴naphthyl-oate)-3α-Larabinofuranosyl-(2′-1″)-α-L-arabinofuranoside

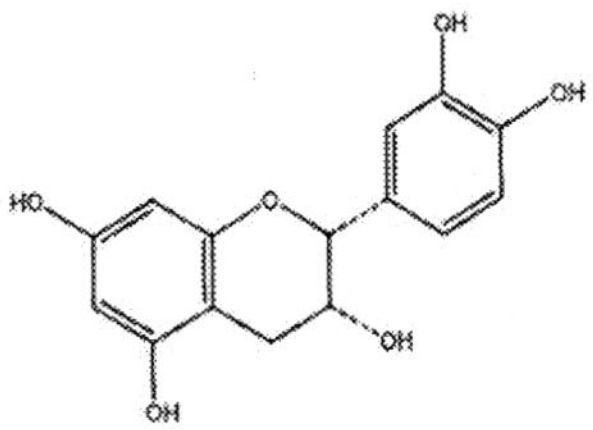

(-)- Epicatechin

CH_3-$(CH_2)_{25}$-CH_2-COOH

Octacosanoic acid

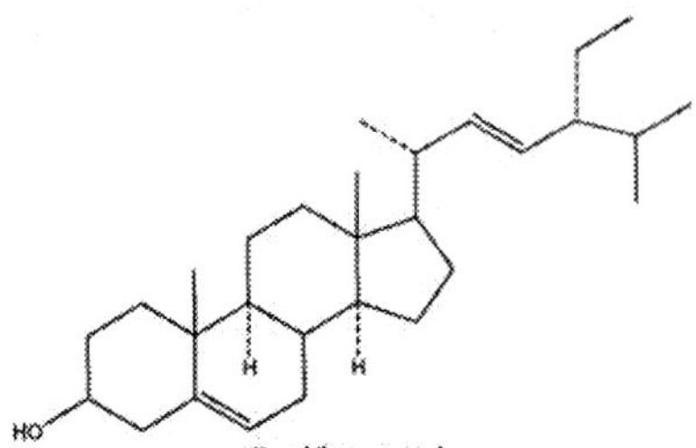

Poriferasterol

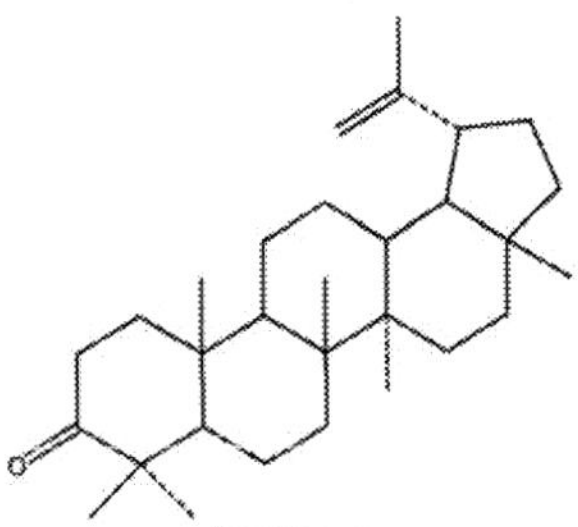

Lupenone

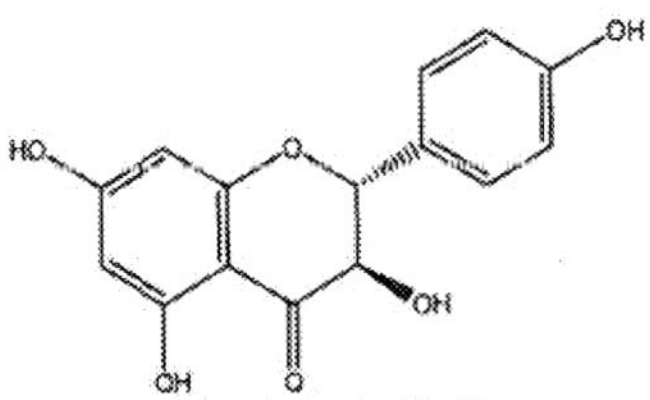

Dihydrokaempferol

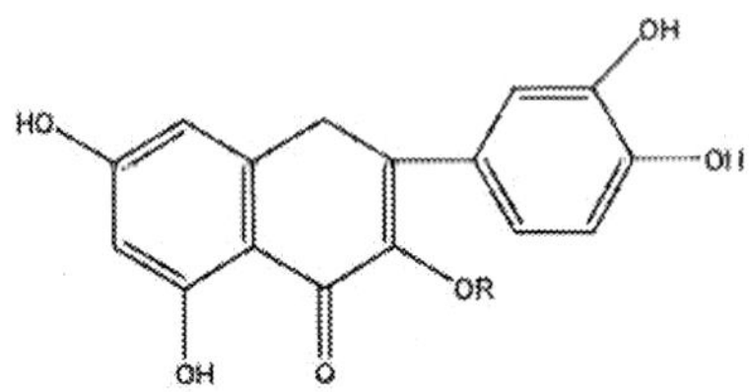

Quercetin R=H
Quercitrin R=Rhamnosyl
Quercimeritrin R=Glucosyl

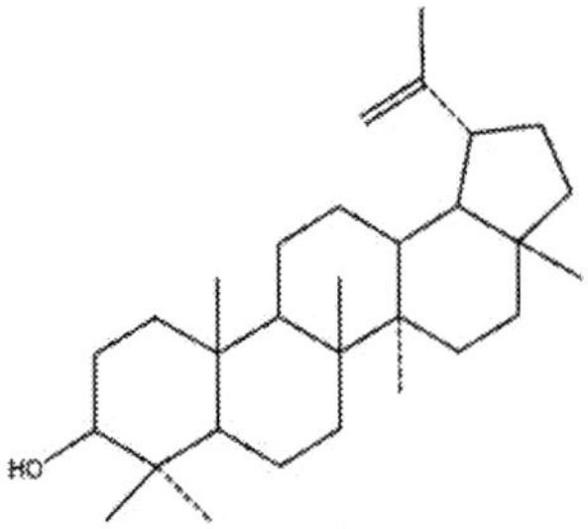

Lupeol

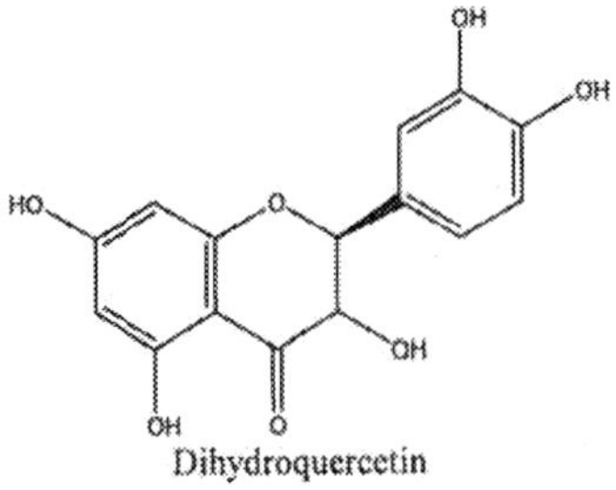

Dihydroquercetin

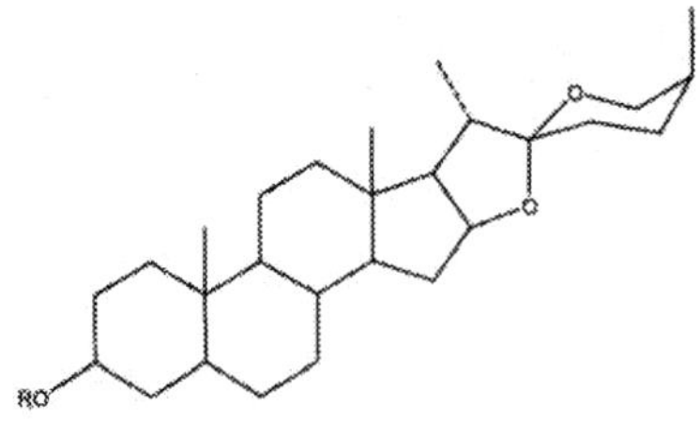

R=Glu(4→1) Rham(2→1) Glu Shatavarin IV

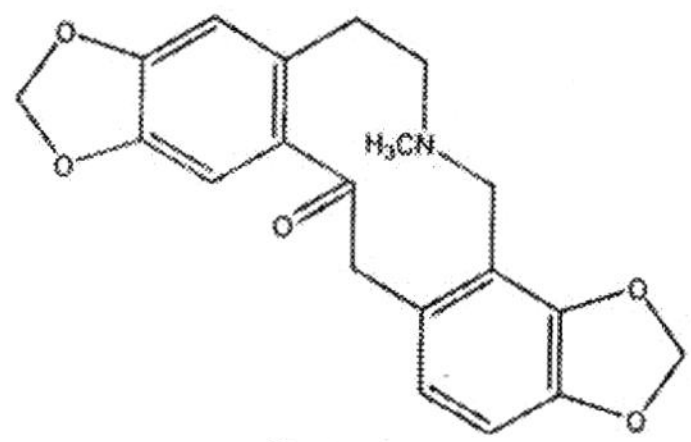

Protopine

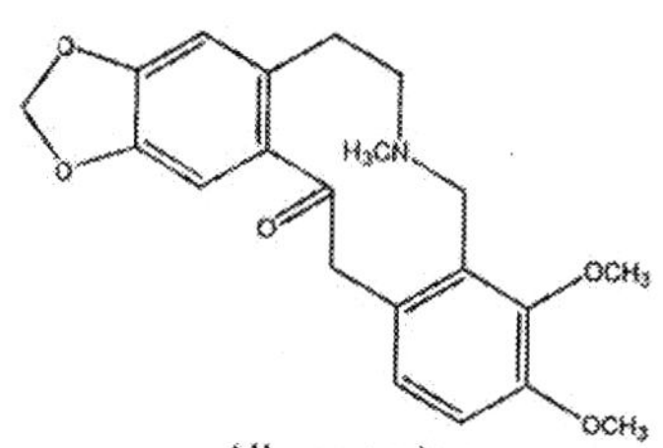

Allocryptopine

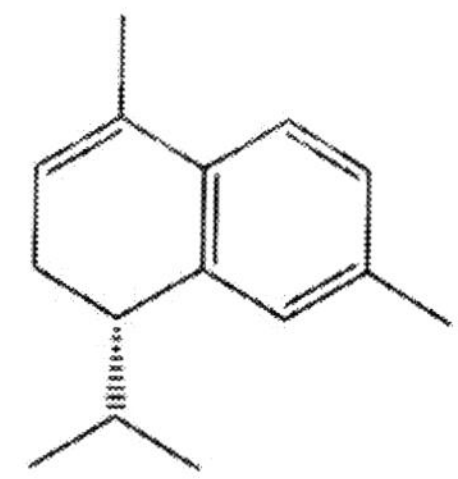
α-Calacorene

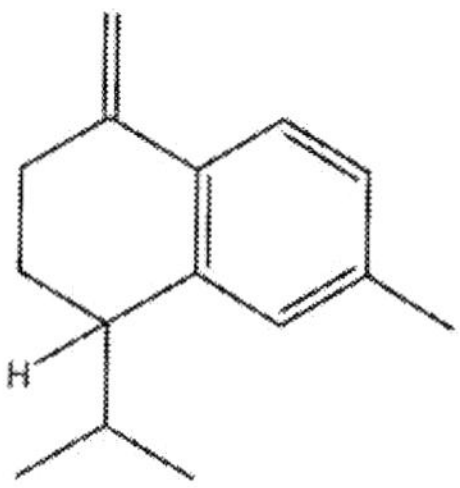

β-Calacorene

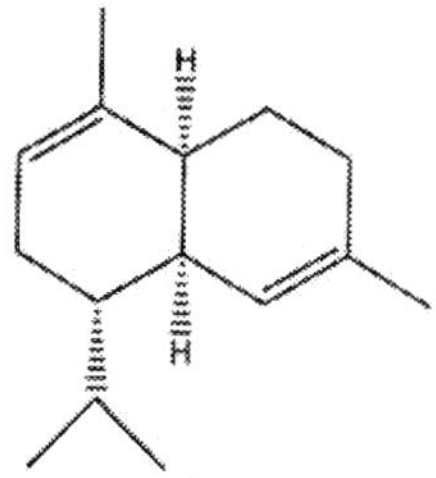

α-Murrolene

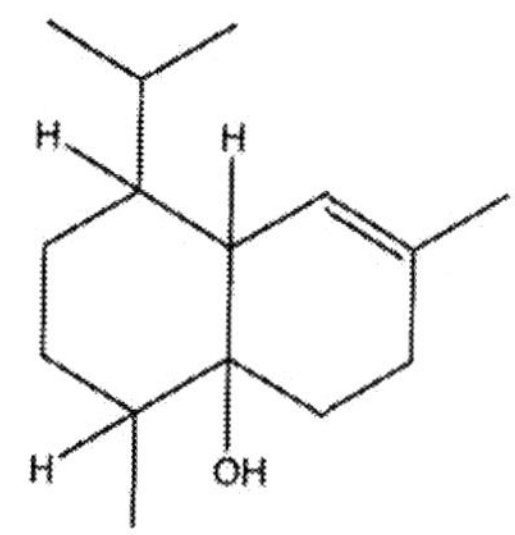

Cubenol

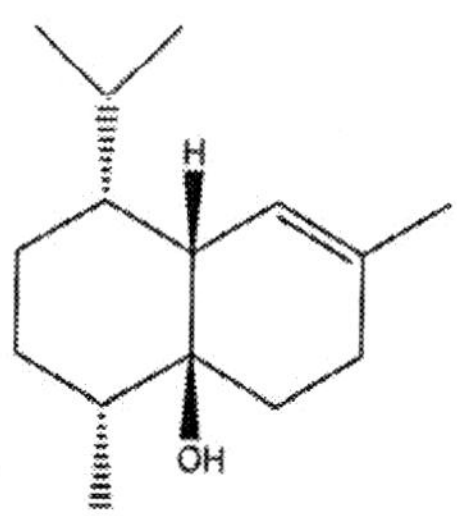

Epi-Cubenol

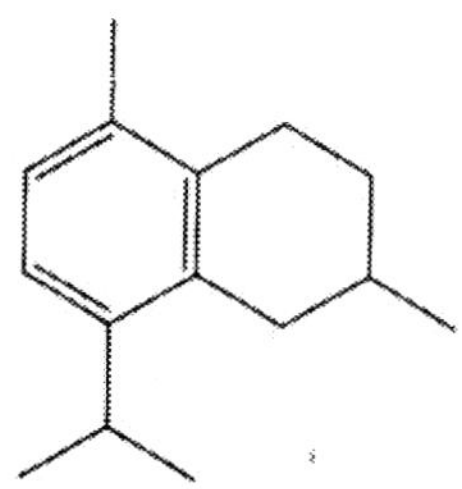
Calamene

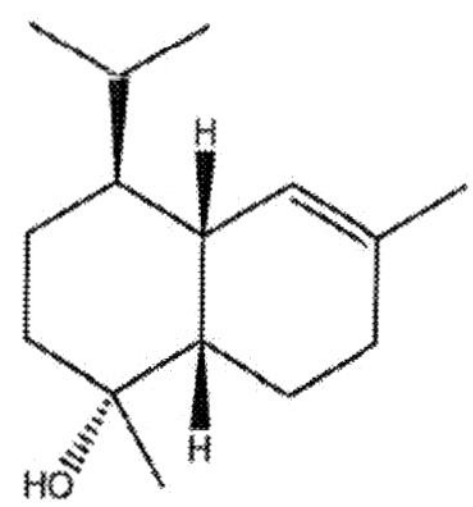

Torreyol

6

CHAPTER

Analysis of Essential Oils from Some Plants

INTRODUCTION

The essential oils or volatile oils or aromatic oils are odoriferous substances, mostly constituted of monoterpenoids (made of two isoperene units), oxygenated monoterpenoids, sesquiterpenoids (made of three isoprene units), oxygenated sesquiterpenoids and are extensively distributed throughout the plant kingdom. The essential oils are normally liquid, but can also be solid (*Orris*) or semi-solid according to temperature (rose) *Illustrated Encyclopedia of Essential Oils***. These occur in special cells, glands or ducts situated occasionally in different parts of plants (Gunther, 1948; Ortiz and Robillos, 1980). Some examples of their occurrence in different parts are given below.

Name of plant part	Source (Common name)	Botanical name
Bark	Cinnamon	*Cinnamomum schaeffer*
Buds	Cloves	*Eugenia caryophyllus*
Flowers	Rose	*Rosa damascena*
Peels	Orange	*Citrus aurantium*
Leaves	Eucalyptus	*Eucalyptus globulus*
Seeds	Cardamom	*Elettaria cadamomum*
Twigs and Leaves	Tagetes	*Tagetes minuta*
Roots	Ginger	*Zingiber officinalis*
Roots and Rhizomes	Indian Valerian, Jatamansi	*Valeriana wallichii*
Wood	Sandal	*Santalum album*
	Agar	*Aquilaria agallocha*
	Deodar	*Cedrus deodara*
	Pine	*Pinus roxburghii*

The oils are usually present in very small amounts and consist only a minute fraction of whole plant material. The oils are produced during some metabolic processes of the plant and are secreted or excreted as odoriferous by products. The essential oils are found to occur in about 65 families and their occurrence was more abundant in families like *Burseraceae, Leguminosae, Myrtaceae, Rosaceae, Rutaceae, Simacroubaceae, Umbelliferae,* etc., (Krishna & Badhwar, 1948, 1949, 1950, 1952, 1953). Some other essential oil bearing families are as given below.

S. No.	Name of family
1.	*Anacardiaceae*
2.	*Aquifoliaceae*
3.	*Begoniaceae*
4.	*Caricaceae*
5.	*Celastraceae*
6.	*Combretaceae*
7.	*Crassulaceae*
8.	*Cucurbitaceae*
9.	*Dilleniaceae*
10.	*Hamamelidaceae*
11.	*Hippocrateaceae*
12.	*Icacinaceae*
13.	*Lecythidaceae*
14.	*Lythraceae*
15.	*Meliaceae*
16.	*Moringaceae*
17.	*Olacaceae*
18.	*Passifloraceae*
19.	*Rhamnaceae*
20.	*Sabiaceae*
21.	*Sapindaceae*
22.	*Saxifragaceae*
23.	*Sonneratiaceae*
24.	*Vitaceae*

Essential oils are normally soluble in pure alcohol, fats and oils, but not in water and evaporate on exposure to air leaving no oily residue behind (*Illustrated Encyclopedia of Essential Oils*). The chemical constituents of essential oils may be divided into four groups which are specific of the bulk of essential oils and are mentioned below (Gunther, 1948).

1. Monoterpenoids, oxygenated monoterpenoids, sesquiterpenoids, oxygenated sesquiterpenoids and diterpenoids (not common) derived from isoprene units.
2. Straight chain carbon compounds
3. Benzene derivatives
4. Miscellaneous compounds

ISOLATION OF ESSENTIAL OILS

The essential oils may be isolated from aromatic materials of the plant origin by using any of the following methods. The suitability of the method depends on the nature of the oil in the material. The essential oils are usually isolated by hydrodistillation, but many methods are known which are used from lab to commercial scale. These methods are as follows (Gunther, 1948).

1. Hydrodistillation
   (a) Water distillation
   (b) Water and steam distillation
   (c) Direct steam distillation
2. Expression
3. Solvent extraction
4. Enfleurage

Hydrodistillation

Distillation may be defined as "the separation of constituents of a mixture of two or more liquids by virtue of the difference in their vapour pressure" (Stephen, 1940). The essential oils are isolated from the aromatic plant material normally by hydrodistillation and in this process essential oil and water forms a two-phase liquid. The oil present in the plant material is made to leave the surface by boiling it in the presence of water and converting it into vapour. The heat energy has to be applied for changing any liquid into vapour and called the latent heat of vaporisation. Reversely, if any vapour is changed back into liquid, its energy is reduced and the latent heat is surrendered and given out.

Also, all compounds with determinable boiling points limit fast moving vapour molecules from their surface. If two mutually insoluble compounds like oil and water are present in the same vapour space, each will exert the pressure appropriate to the concentration of its own molecules, as though the other were not there. The total pressure exerted by the mixed vapour then becomes the sum of the partial pressures exerted by each constituent present. It follows that a liquid mixture will boil when its temperature is raised to the point where the combined vapour pressure of its components becomes equal to the surrounding pressure.

The basic principle of this hydrodistillation, consists in bringing the two mutually insoluble liquids into contact at a temperature very close to the boiling point of the more volatile of the two, the water. Then the addition of small amount of vapour pressure from the oil will make the combined pressure equal to the surroundings pressure. The mixture will then boil and the oil will vaporise into the passing steam as fast as its requirement of latent heat can be applied. It is the rate of application of this latent heat that governs the speed and efficiency of the distillation. Distillation is believed to be complete when no more oil appears in the condensate. The basis of hydrodistillation normally may be represented as follows.

The ratio between the weights of two vapour components (Steam and oil), and therefore of two liquids (water and oil in case of essential oils) in the distillate (condensate), is denoted by the ratio of their vapour pressure multiplied by the ratio of their molecular weight.

$$\frac{\text{Weight of water}}{\text{Weight of oil}} = \frac{\text{Vapour pressure of water}}{\text{Vapour pressure of oil}} \times \frac{\text{Molecular weight of water}}{\text{Molecular weight of oil (average)}}$$

Distillation is affected by the variation of pressure inside still. The reduced pressure causes the decrease in the boiling temperature of the liquid. The decrease in temprature leads to rapid decrease of oil pressure than of water, thereby causing increase in the quantity of water in the distillate. The condenser efficiency increases due to low temperature of vapour. The reduced pressure distillation is beneficial for essential oil due to absence of decomposition and heat. But, the vaporisation ability of compounds having high boiling temperature is considerably reduced. The excess pressure distillation causes the decomposition of the oil (Datta and Borah, 1986).

General Methods of Distillation

The essential oils can be isolated by three types of distillation given as:

(a) Water distillation
(b) Water and steam distillation
(c) Steam distillation

Water Distillation

The aromatic material from which the essential oil is to be isolated remains in direct contact with water and may float inside the retort depending on the density of the material. The water is boiled using a suitable method. The essential oil from the chipped sandal wood, agarwood, rose flowers and other materials may be isolated using this method. The Clevenger type apparatus is used for the isolation of essential oil on laboratory scale.

Water and Steam Distillation

The aromatic plant material is placed on a perforated grid situated at a suitable distance above the bottom of the still. The empty part below the gird is filled with water to a convenient level. The low pressure wet steam is generated on heating the water either with direct fire or some other convenient device, which rises through the plant material. The advantage of this method over water distillation is that plant material is not in the direct contact with boiling water, thereby preventing the scope of direct hydrolysis and saturated steam is available for the removal of essential oil. The water distillation and this method are used for small scale distillation of the essential oil. The practice of returning distilled water to vat for re-distillation is known as cohobation.

Direct Steam Distillation

The steam distillation method is similar with water and steam distillation, except that water is not placed inside the still. The steam generated outside the still is inserted into the still below the aromatic material charge, through a perforated steam coil. The steam may be saturated or superheated and ascends through the plant material. The method is used for large scale distillation of the oil. The retort or still, the condenser, and the receiver are the common major components of all the distillation plants. The boiler or steam generator constitutes the fourth major part of steam distillation method (Datta and Borah, 1986).

Expression

A number of compounds of essential oils from *Citrus* peels are thermally labile and may be degraded by heat on exposure to steam distillation. The cold expression method is therefore used almost wholly for the production of oils contained in the rinds of fruits, such as orange, lemon, bergamot and others. It was initially started as household process, but the industrial development has taken place with the passage of time. The fruits peel is squeezed, lacerated or abraded to rupture the oil cells in the exocarp for releasing the essential oil in the cold expression method. The process of expression normally involves squeezing the material of great pressures in order to press out oils; the process is carried out either in manual presses or crushers on a cottage industry basis or in huge mechanical presses on industrial scale. The oil may be purified suitably at a later stage. The modern expression methods are based on two traditional methods of Italy.

Solvent Extraction

The solvent extraction process is used for the isolation of essential oils from the aromatic plant materials consisting of thermolabile compounds, containing low yield of oil, and to isolate non-volatile compounds required for fixative purposes (preparation of resinoids from exudates). It is an industrial process and normally uses the low boiling solvents like hexane, petroleum ether, ether or other suitable solvent for the extraction of plant material. The solvent extraction can be carried out by a selectively chosen volatile solvent by feeding into the top of the vessel and allowed to steep through the aromatic plant material. The solvent, on penetration into the plant material, dissolves the aromatic compounds along with waxes, albuminous and colouring matters. After the extraction, the solvent is distilled off, leaving a semi-solid which consists of oils and insoluble waxes. The waxes are removed by the addition of alcohol followed by filtration or by freezing and thereafter the alcohol is eliminated. The resulting absolute is highly concentrated form of perfume. Solvent extraction is used in making the concrete resinoids and absolutes.

Some latest solvent extraction methods for the isolation of essential oil from aromatic plant material are given below (Jirovetz and Buchbauer, 2005).

(a) Supercritical fluid extraction (SFE)
(b) Extraction with other liquefied gas solvents
(c) Extraction with accelerated solvent (ASE)
(d) Solvent extraction with sonication
(e) Microwave-assisted extraction (MAE)
(f) Solvent-free microwave extraction
(g) Solid phase microwave extraction
(h) Continuous subcritical water extraction (CSWE)

The detailed descriptions of the above methods have been described in the essential oil book (Jirovetz and Buchbauer, 2005).

Enfleurage

Enfleurage or extraction with odourless fats is carried to get the fragrant compounds exuded by plants. Normally, the process uses cold fat (cold enfleurage) but hot fat may also be used (hot enfleurage). The process is used on small scale as it is a labour intensive and costly process. The essential oil from jasmine and other delicate flowers is isolated using the process.

FUNCTIONS OF ESSENTIAL OILS TO PLANTS

The presence of essential oil in the plants has been considered a lure to insect pollinators of the flowers, as moderator in the intracellular oxidation to protect against the action of atmospheric agents and as a safe guard against parasites. They also exercise a definite action on the transpiration of plants. Monoterpenes present in the essential oils provide material for the synthesis of physiologically important pigments. Specific monoterpenes are claimed to act growth, heat and transpiration regulators and as participants in photosynthesis. Certain components of the essential oils are also useful in seed germination and seedling growth inhibition (Jirovetz and Buchbouer, 2005).

GENERAL USES OF ESSENTIAL OILS

Essential oils are used mainly in perfumery, fragrance and medicinal purposes in industries.

(a) Soap and cosmetics
(b) Pharmaceuticals
(c) Confectionary
(d) Attars, agarbathies, incense, scented tobacco, etc.
(e) Turpentine oil is used on a large scale in the manufacture of paint and varnishes.
(f) Aromatherapy

ANALYSIS OF ESSENTIAL OILS

The essential oils are a complex mixture of closely related compounds and are generally not possible to isolate in pure form by column chromatography. The compounds in oils are best identified by gas chromatography-mass spectrometry (Marriot et al., 2001). The details about the analysis have been described in a number of standard books.

Isolation and Characterisation of Essential oils from Some Plants

(A) *Eucalyptus globulus* ssp. *bicostata* (Maiden, Blakely and J. Simm)

1. **Name of the plant species:** *Eucalyptus globulus* ssp. *bicostata*
   (a) **Family:** Myrtaceae
   (b) **Parts used for the isolation of essential oil:** Leaves
2. **Method of the isolation of essential oil:** The essential oil was isolated by hydro-distillation of the chopped leaves in a Clevenger-type apparatus till the complete removal of the essential oil, followed by removal of the essential oil from the liquid distillate by extraction with diethyl ether, drying of ethereal layer over anhydrous sodium sulphate, and removal of the solvent on a gently heated water bath.

3. **Research contribution of Dayal and Ayyar:**
   (a) **Essential oil yield (Moisture free basis):** 6.58%
4. **Method of characterisation:** The analysis of the oil was carried out using gas liquid chromatography (GLC) using a butanediol succinate polyester 20% on celite, 1.2 m × 0.3 cm column and FID detector, N_2, 40 ml/min; 160°C. The other experimental conditions of the analysis have been described. (Dayal and Ayyar, 1986).
5. **Number of compounds identified:** 5

   Five compounds namely α-pinene (4.0%), 1,8-cineole (73.09%), linalool (0.09%). α-terpineol (0.7%), and bicostol (15.70%), constituting (93.58%) of the oil were identified, while 6.32% of the essential remained unidentified (Dayal and Ayyar, 1986).
6. **Chief constituent:** 1,8-Cineole (73.09%)
7. **Major compounds (>20%):** 1,8-Cineole (73.09%)
8. **Structures of major constituents:** (Given at the end of this chapter)
9. **Conclusion:** This is the first report of the isolation and characterisation of essential oil from *Eucalyptus globulus* ssp. *bicostata* leaves growing in Dehra Dun, India and 1,8-cineole was found to be the chief constituent. The study indicates that the leaves can be used as an excellent source of medicinal oil (better than *E. globulus)* due to presence of 73.09% of the 1,8-cineole (Dayal and Ayyar, 1986, 1986a). Bicostol was found to be a new compound.

(B) *Eucalyptus* hybrid (Mainly *E. umbellata* Domin syn. *E. tereticornis* sm.)

1. **Name of the plant species:** *Eucalyptus* hybrid
   (a) **Family:** Myrtaceae
   (b) **Parts used for the isolation of essential oil:** Leaves
2. **Method of isolation:** The essential oil from the chopped leaves was isolated by hydrodistillation using a Clevenger-type apparatus followed by other experimental steps as given under A(2).
3. **Research contribution of Dayal, Parcha and Bartley:**
   (i) **Essential oil yield (Moisture free basis):** 1.82%
4. **Method of characterisation:** The analysis of essential oil was carried out by its GC-MS on a VGMD 800 bench top instrument (Fisons Instruments Manchester, UK) equipped with a fused silica capillary column (J & W DB 5, 60 M 0.25 mm, coating thickness 0.25 μm). The other experimental conditions of the analysis have been described (Dayal et al., 1997).

5. **Number of compounds identified:** Nine
   This investigation on the analysis of oil indicated the presence of 80 compounds out of which 71 remained unidentified. The identified nine compounds were found to be α-pinene (8.34%),. β-pinene (4.43%), myrcene (0.24%), terpinene or phellandrene (12.73%), p-cymene (7.25%), limonene (1.97%), 1,8-cineole (52.11%), borneol (0.22%), and α-terpineol (0.82%) constituting 88.11% of oil while 71 compounds (11.89%) remained unidentified (Dayal et al., 1997).
6. **Chief constituent of the oil:** 1,8-Cineole
7. **Major compounds (>15%):** Cineole (52.11%)
8. **Structure of major constituents:** (Given at the end of this chapter)
9. **Conclusion:** The analysis of essential oil by GC-MS indicated the presence of 80 compounds and this number was not reported earlier. 1,8-Cineole was found to be the chief constituent.

(C) *Eucalyptus* species (Six)

1. **Name of the plant species:** FRI-4(*Eucalyptus tereticornis x Eucalyptus camaldulensis)*, FRI-5(*Eucalyptus camaldulensis x Eucalyptus tereticornis), Eucalyptus tereticornis, Eucalyptus camaldulensis, Eucalyptus robusta,* and *Eucalyptus deglupta*
   - **(a) Family:** Myrtaceae
   - **(b) Parts used for the isolation of essential oil:** Leaves
2. **Method of isolation of essential oil:** The essential oil from the chopped leaves of all the *Eucalyptus* species was isolated by hydrodistillation in a Clevenger-type apparatus followed by further work up as given under A(2).
3. **Research contribution of Dayal and Maheshwari:**
   - **(a) Essential oil yield (Moisture free Basis):**

E. camaldulensis	1.37%
E. *tereticornis*	0.95%
FRI-4	1.84%
FRI-5	1.17%
E. robusta	1.70%
E. deglupta	0.26%

4. **Method of characterisation:** The chemical constituents present in all the five essential oils were identified by their Gas Liquid Chromatography (GLC) over Hewlett Packard Gas Chromatograph Model 5835A equipped with Carbowax (10% 20M) column, FID detector and a printer plotter using programming (Dayal and Maheshwari, 1985).

5. **Number of Compounds Identified:** 15

   Analysis of the essential oil of all *Eucalyptus* species using GLC showed the presence of 23 compounds of which 15 compounds were identified in all the species. The identified compounds were α-pinene, camphene, β-pinene, phellandrene, limonene, 1,8-cineole, γ-terpinene, p-cymene, citronellal, linalool, terpin-1-ene-4-ol, citronellal acetate, borneol, α-terpineol, and piperitone (Dayal and Maheshwari, 1985).

6. **Chief constituents:**

E. camaldulensis	Terpin-1-ene-4-ol (10.50%)
E. tereticornis	Limonene (11.71%)
FRI-4	α-Terpineol (8.81%)
FRI-5	α-Pinene (13.18%)
E. robusta	Piperitone (11.37%)
E. deglupta	p-Cymene (7.90%)

7. **Major constituents (>8%):**

E. camaldulensis	Terpin-1-ene-4-ol (10.50%)
	β-Pinene (10.29%)
E. tereticornis	Limonene (11.71%)
FRI-4	α-terpineol (8.81%)
	1,8-Cineole (8.15%)
FRI-5	α-Pinene (13.18%), *p*-Cymene (8.65%)
E. robusta	Piperitone (11.37%), α-Terpineol (9.99%)
E. deglupta	p-Cymene (7.90%)

8. **Structures of major constituents:** (Given at the end of this chapter)
9. **Conclusion:** This was the first report on the isolation and characterisation of the fifteen compounds in the essential oils of FRI-4 and FRI-5. The analysis of other *Eucalyptus* species oil was also carried out. Physicochemical properties of the oils were also determined.

(D) *Pinus insularis* Endl.syn. *Pinus khasya* Royle and *Pinus merkusii* Jungh. & De Vriese

1. **Name of the plant species:** *Pinus insularis* and *Pinus merkusii*
   - (a) **Family:** Pinaceae
   - (b) **Parts used for the isolation of essential oil:** Oleoresin
2. **Method of isolation of essential oil:** The essential oil from both the oleoresins was isolated by hydrodistillation in a Clevenger –type apparatus followed by other experimental work up as given under A(2).

3. **Research contribution of Dayal:**

   (a) **Essential oil yield:** *Pinus insularis* – 21%, *Pinus merkusii* -23%

4. **Method of characterisation:** Both the essential oils were characterised by their Gas Liquid Chromatography(GLC) on a CIC Gas Chromatography equipped with Flame Ionisation Detector (F.I.D); using 20% Butanediol succinate polyester on celite column (1.2 m × 0.3 cm), separately. The other conditions of the analysis have also been given (Dayal, 1986).

5. **Number of compounds identified: (*P. insularis* -3, *P. merkusii* -2)**
   Three compounds, namely, α-pinene, β-pinene and longifolene were identified in the oil of *P. insularis* and α-pinene and delta-3-carene in the oil of *P. merkusii* (Dayal, 1986). Three compounds (8.46%) in *Pinus insularis* and one compound (3.60%) in *Pinus merkusii* remained unidentified.

6. **Chief constituent:**
   P. insularis - α-pinene (59.36%),
   Pinus merkusii – α-pinene (57.05%)

7. **Major constituent(>10%):**
   Pinus insularis α-pinene (59.36%) and β-pinene (31.20%);
   P. merkusii α-pinene (57.05%) and delta-3-carene (39.35%).

8. **Structures of major constituents:** (Given at the end of this chapter)

9. **Conclusion:** The essential oil isolated from the oleoresin of *P. insularis* and *P.merkusii* were found to be rich in α-pinene, β-pinene and α-pinene, delta 3-carene, respectively and thus these two oleoresins can be exploited on a commercial scale for the isolation of oil and rosin. This was the first report on the chemical analysis of the essential oil isolated from the oleoresin of *P. merkusii* (collected from Waloon Lohit Division) and *P. insularis* (collected from Bunder Dev Division) of Arunachal Pradesh, India.

(E) *Vitex negundo* Linn

1. **Name of the plant species:** *Vitex negundo*

   (a) **Family:** Verbenaceae

   (b) **Parts used for the isolation of essential oil:** Leaves, Flowering twigs

2. **Method of isolation:** The essential oil from the chopped shade dried leaves and the fresh flowering twigs was isolated by hydrodistillation using Clevenger-type apparatus for 3 hrs and 4 hrs, respectively followed by further work up as described under A(2).

3. **Research contribution of Singh, Dayal, Bartley and Bhandari:**
   (a) **Essential oil yield (moisture free basis):** Leaves 0.12%, flowering twigs 0.08%
4. **Method of characterisation:** The identification of different constituents in both the essential oils was performed using gas chromatography – mass spectrometry (GC-MS) on a VGMD 800 bench top instrument (Fisons Instruments) equipped with a fused capillary column (J & W DB-5, 60 m × 0.25 mm, coating thickness 0.25 µm). The other experimental details of examination have also been described (Singh et al., 1999; Singh et al., 2000).
5. **Number of compounds identified:** Leaves-35, Flowering twigs-28
   GC-MS analysis of the leaves and flowering twigs oils indicated the presence of 66 and 94 compounds of which 35 and 28 compounds were, characterised respectively. The identified leaves compounds (74.96%) were found to be thujene, α-pinene, sabinene, 1-oceten-3-ol, β-pinene, myrcene, delta-3-carene, p-cymene, limonene, β-phellandrene, gama-terpinene, dihydromyrcenol, sabinene hydrate, linalool, amyl isovalerate, 1-methyl- 4-isopropylcycohex-2-en-1-ol, nonaonol, 4-terpinene, α-terpineol, 3-methyl-6-isopropylcyclohex-2-en-1-ol, carveol, eugenol, α-gurgenene, β-caryophyllene, isocaryophyllene, humulene, aromadendrene, viridiflorene, delta-cadinene, elemol, caryophyllene oxide, virdiflorol, globulol, eudesmol, and gama-eudesmol (Singh et al., 1999), while the flowering twigs contained (57.53%) of the identified compounds, namely, 3-hexen-1-ol, 2-methylcyclopentenol, thujene, α-pinene, sabinene, 1-oceten-3-ol, β-pinene, myrcene, delta-3-carene, p-cymene, limonene, β-phellendrene, 1,8-cineole, gama-terpinene, dihydromyrcenol, sabinene hydrate, linalool, nonanol, 4-terpineol, α-terpineol, eugenol, caryophyllene, β-farnesene, aromadendrene, elemol, caryophyllene oxide, virdiflorol, and globulol (Singh et al., 2000).
6. **Chief constituent:** Leaves—Virdiflorol (19.55%); Flowering twigs—Virdiflorol (26.55%)
7. **Major constituents (> 10%):** Leaves Virdiflorol (19.55%); β-Caryophyllene (16.59%); and Sabinene (12.07%); Flowering twigs Virdiflorol (26.52%) and β-Caryophyllene (13.20%)
8. **Structures of major constituents:** (Given at the end of this chapter).
9. **Conclusion:** Thirty five and twenty eight compounds constituting 74.96% and 57.53% of the essential oil of leaves and flowering twigs were identified, respectively. The percentage of virdiflorol was found to be more in the flowering twigs than in the leaves and thus can be a better source of virdiflorol. The essential oil (0.062%) of the leaves exhibited 100% mortality against stored grain pest, augoumois grain moth, *Sitotroga cerealella* infesting wheat seeds and the new adult emergence was completely prevented by 0.125% solution of the oil and thus may be of use to control them. Virdiflorol exhibited significant antifeedant activity

against rice weevil, *Sitophilus orygae* and ovipositional activity against black gram beetle, *Callosobruchus chinensis* (Singh et al., 2002).

(F) *Shorea robusta* Gaerti

1. **Name of the plant species:** *Shorea robusta*
   - **(a) Family:** Dipterocarpaceae
   - **(b) Parts used for the isolation of essential oil:** Heartwood, Resin, Leaves, Bast
2. **Method of isolation:** The essential oil from the heartwood and resin was isolated by hydrodistillation of their respective petroleum ether extract, while leaves and bast yielded essential oils by hydrodistillation of the chopped material in a Clevenger-type apparatus for 6 hrs each followed by removal of the essential oil from the liquid distillate as given under A(2).
3. **Research contribution by Kaur, Varshney, Dayal and Bartley:**
   - **(a) Essential oil yield (on moisture free basis):** Heartwood 0.01%; Resin 0.12%; Leaves 0.08%; Bast 0.015%
4. **Method of characterisation:** The identification of different compounds present in the above cited essential oils was carried out using gas chromatography-mass spectrometry (GC-MS) on a VGMD 800 bench top instrument (Fisons Instrument) equipped with fused silica capillary column (J & W DB-5, 60 m × 0.25 mm, coating thickness 0.25 μm) for the essential oil of heartwood, resin and leaves while HP 5890-Trio-1, equipped with a fused silica capillary column (BPX-5, 30 × 0.32 mm, 0.25 μm thickness) was used for the analysis of bast oil.
5. **Number of compounds identified:** Leaves 8; Heartwood 9; Bast 9; Resin 17

   GC-MS analysis of the essential oils from leaves (LO), heartwood (HO), bast (BO), and resin (RO), revealed the presence of 20, 24, 28, and 37 compounds of which 8, 9, 9, and 17 compounds representing 98.30%, 80.40%, 48.80%, and 78.40% of the oil, respectively, were identified. The leaves oil was found to contain α-copaene, β-elemene, β-caryophyllene, α-caryophyllene, germacrene-D, β-farnesene, spathulenol, and gama-eudesmol (Kaur et al., 2001), while the HO consisted of α-cubebene, α-copaene, β-caryophyllene, α-caryophyllene, germacrene-D, gama-cadinene, virdiflorene, globulol, and caryophyllene oxide (Kaur et al., 2001a). The identified chemical constituents of RO were α-pinene, camphene, β-pinene, p-cymene, limonene, α-cubebene, aciphylene, α-copaene, β-elemene, β-caryophyllene, α-caryophyllene, alloaromadendrene, gama-muurolene, germacrene-D, gama-cadinene, globulol, and caryophyllene oxide (Kaur et al., 2001a), while BO contained α-terpineol, α-copaene, α-elemene, gama-muurolene, gama-cadinene, virdiflorene, globulol, τ-cadinol, and α-cadinol (Kaur et al., 2003; Varshney et al., 2005).

6. **Chief constituents of different oils:** Leaves: α-Caryophyllene (54.55%); Heartwood: Germacrene-D (31.79%); Resin: Germacrene-D (29.57%); Bast: τ–Cadinol (16.75%).

7. **Major constituents (> 10%):** Leaves α-Caryophyllene (54.55%), β-Caryophyllene (28.27%), β-farnesene (12.62%); Heartwood: Germacrene-D (31.79%), β-Caryophyllene (24.36%), α-Copaene (10.28%); Resin: Germacrene-D (29.75%), α-Copaene (16.32%), Globulol (10.22%); Bast: τ-Cadinol (16.75%), α-Cadinol (16.45%).

8. **Structures of major constituents:** (Given at the end of this chapter).

9. **Conclusion:** This was the first report of the isolation and characterisation of essential oils from the leaves (LO), heartwood (HO), resin(RO), and bast (BO) of *Shorea robusta* with the presence of α-caryophyllene, germacrene-D, gemacrene-D, and τ-cadinol as the chief constituents, respectively. LO, HO, BO and ethereal extract of the resin (EER) exhibited the electrophysiological activity in the female beetle, while LO and BO in the male beetle, *Hoplocerambyx spinicornis* NEWM; of the injurious heartwood borer of *Shorea robusta* on bioassay with electroantennograph (EAG). BO alone elicited the attractant activity to both the sexes in the wind-tunnel studies. BO was identified by GC/MS and the correlation of the constituents was also studied (Varshney et al., 2005).

(G) *Cephalotaxus harringtonia* (Knight ex Forbes) Koch var. *harringtonia*

1. **Name of plant species:** *Cephalotaxus harringtonia*
   - **(i) Family:** Cephalotaxaceae
   - **(ii) Parts used for the isolation of essential oil:** Needles and Twigs
2. **Method of isolation of essential oil:** By hydrodistillation of the chopped plant material for 6 hrs each in a Clevenger-type apparatus followed by removal of essential oil from the distillate as given under A(2).
3. **Research contribution of Mendiratta, Dayal and Bartley:**
   - **(a) Essential oil yield (Moisture free basis):** Needles 0.028%
     Twigs 0.011%
4. **Method of characterisation:** The oils were analysed separately by Gas Chromatography-Mass-Spectrometry (GC/MS) on a VG 800 bench top instrument (Fisons Instruments) equipped with fused silica capillary column (J & W DB-5, 60 m × 0.25 mm, coating thickness 0.25 μm). The experimental conditions of the oil analysis have been explained (Mendiratta et al., 2005).
5. **Total number of compounds identified:** Needles 17
   Twigs 17

A total number of 17 compounds were characterised from each, needles and twigs namely γ-terpinene, terpinen-4-ol, α-terpineol, β-copanene, β-elemene, α-carophyllene, α-humulene, γ-muurolene, germacrene-D, α-muurolene, α-bulnesene, γ -cadinene, globulol, caryphyllene oxide, α-bisabolol, τ-cadinol, and β- eudesmol (Mendiratta et al., 2005).

6. **Chief constituent:** Needles—β-caryophyllene (27.9%); Twigs—β-caryophyllene (31.3%)
7. **Major constituents (> 12%):** Needles: β-caryophyllene (27.9%);
   Twigs: β-caryophyllene (31.3%);
   germacrene-D (13.0%)
8. **Structures of major constituents:** Given at the end of this chapter
9. **Conclusion:** Five sesquiterpenes—namely, β-caryophyllene (27.9–31.3%), –α-humulene (11.4–10.6%), germacrene-D (9.50–13.0%), γ-cadinene (8.90–5.50%) and β-elemene (8.30-10.90%) were identified as the main constituents in needles and twigs oil by GC-FID. β-Caryophyllene was found to be the chief constituent in both the oils. The analyses of these essential oils were reported for the first time from *Cephalotaxus harringtonia* growing in India (Mendiratta et al., 2005).

(H) *Tagetes minuta* Linn. syn. *T. glandulifera* Schrank

1. **Name of the plant species:** *Tagetes minuta*
   - **(a) Family:** Asteraceae, Compositae
   - **(b) Parts used for the isolation of essential oil:** Aerial parts (Leaves, Blossom and very small stalks)
2. **Method of isolation:** The essential oil from the chopped, air dried aerial parts collected from Dehra Dun region of Uttarakhand, India was isolated by hydrodistillation in a Clevenger-type of apparatus for 6 hrs followed by further experimental procedure as given under A(2).
3. **Research contribution of Kumar, Dayal and Sharma:**
   - **(a) Essential oil yield (Moisture free basis):** 0.80%
4. **Method of characterisation:** The essential oil of the aerial parts was analysed by gas chromatography-mass spectrometry and gas chromatography was used for the determination of the amount of individual constituents of oil. The GC-MS analysis was carried out on a Agilent 5975 Inert Mass Selective Detector interfaced with an Agilent (HP 6890) GC system fitted with capillary column of HP-5MS

(5% phenyl methyl siloxane, 30 m × 250 μm × 0.25 μm film thickness). The other experimental details of the analysis have also been given (Kumar et al., 2009).

5. **Number of compounds identified:** Seventeen compounds namely sabinene, o-cymene, limonene, (z)-β-ocimene, E-β-ocimene, dihydrotagetone, p-menth-4-ene-3-one, allo-ocimene, (E)-tagetone, (z)-tagetone, terpinen-4-ol, (z)-ocimenone, (E)-ocimenone, β-caryophyllene, α-humulene, spathulenol, caryophyllene oxide constituting 94.36% of the oil were identified. Forty three compounds (5.64%) remained unidentified (Kumar et al., 2009).
6. **Chief constituent:** Dihydrotagetone (45.74%)
7. **Major constituents (> 10%):** Dihydrotagetone (45.74%), (z)-β-ocimene (22.80%)
8. **Structures of major constituents:** (Given at the end of this chapter)
9. **Conclusion:** Seventeen compounds, identified in the essential oil were found to consists of monoterpenes (29.93%), oxygenated monoterpenes (60.71%), sesquiterpenes (1.61%), oxygenated sesquiterpenes (2.11%). Dihydrotagetone (45.74%, oxygenated monoterpene) was found to be the chief constituent. This is the first report of the isolation and characterisation of essential oil from the aerial parts of *Tagetes minuta* of Dehra Dun region and also of its maximum occurrence in the Indian oil. *Tagetes minuta* occurring in the Dehra Dun region can be utilised for the isolation of dihydrotagetone rich oil.

(I) *Achyranthes aspera* Linn.

1. **Name of the plant species:** *Achyranthes aspera*
   - **(a) Family:** Amaranthaceae
   - **(b) Parts used for the isolation of essential oil:** Leaves
2. **Method of isolation:** The essential oil from the chopped leaves was isolated by hydrodistillation using a Clevenger-type apparatus for 4 hrs followed by other experimental steps as given under A(2).
3. **Research contribution of Rashmi and Dayal:**
   - **(a) Essential oil yield (on moisture free basis):** 0.014%
4. **Method of characterisation:** The essential oil was analysed by GC-MS on (Agilent 19091S-433) equipped with HP-5MS capillary column (30 m × 250 μm × 0.25 mm i.d; film thickness 0.25 mm). The other conditions of the analysis have been described (Rashmi and Dayal, 2007).
5. **Number of compounds identified:** 7

   Seven compounds, namely, p-benzoquinone (trace), hydroquinone (57.70%), spathulenol (1.55%), nerol (1.47%), α-ionone (1.17%), asarone (trace), and eugenol (1.16%) constituting 63.05% of the oil were characterised, while 9 compounds remained unidentified (Rashmi and Dayal, 2007).

6. **Chief constituent:** Hydroquinone (57.70%)
7. **Major compounds (>10%):** Hydroquinone (57.70%)
8. **Structures of major constituents:** (Given at the end of this chapter)
9. **Research findings:** This is the first report on the isolation and characterisation of the essential oil from *Achyranthes aspera* leaves. The study indicated hydroquinone as the major constituent of the oil.

REFERENCES

Datta, D.K. and Borah, U.C. (1986). Essential oil distillation, *Indian Perfumer, 30* (2-3): 345–354.

Dayal, R. and Maheswari, M. L. (1985). Terpenoids of the essential oil of *Eucalyptus. Ind. For.*, **3**, 1077.

Dayal, R. (1986). Studies of oleo-resins of *Pinus merkusii* and *Pinus insularis* from Arunachal Pradesh. *Ind. For.*, 112: 734.

Dayal, R. and Ayyar K.S. (1986). Analysis of medicinal oil from *Eucalyptus globulus* sub sp. *bicostata* leaves, *Planta Medica*, 162.

Dayal, R. and Ayyar K.S. (1986a). *Eucalyptus globulus* sub sp. *bicostata* leaves: A new source of medicinal oil, *Indian Forester*, p. 277.

Dayal, R., Precha, V. and Bartley, J. P. (1997). GC-MS Studies of oil from *Eucalyptus* hybrid leaves, *Indian Perfumer*, 41(4): 60.

Gunther, E. (1984). The Essential Oils Vol. 1–4, D. Van Nostrand Company, Inc. Toronto, New York, London.

Jirovetz, L. and Buchbauer, G. (Editors) (2005). *Processing, Analysis and Application of Essential Oils*, Har Krishan Bhalla & Sons, Dehradun, India, pp. 1–34, 59–90.

Kaur, S., Dayal, R., Varshney, V. K., and Bartley, J. P. (2001). Sesquiterpenoids of *Shorea robusta* leaves. *Indian Perfumer* 45 (3): 185–187.

Kaur, S., Dayal, R., Varshney, V. K., and Bartley, J. P. (2001a). GC-MS analysis of essential oils of heartwood and resin of *Shorea robusta. Planta Medica* 67: 883–886.

Kaur, S., Varshney, V.K. and Dayal, R. (2003). GC – MS analysis of essential oil of *Shorea robusta* Bast. *Journal of Asian Natural Products Research*. V. 5 (3): 231–234.

Krishna, S. and Badhwar, R.L. (1948). Aromatic plants of India, *J. of Scientific & Industrial* Research, Vol. VII (No. 5, 6, 8, 10, 12): pp. 77–79, 101–106, 117–128, 129–144, 9a–9b.

Krishna, S. and Badhwar, R.L. (1949). Aromatic plants of India, *J. of Scientific & Industrial Research*, Vol. 8 (No. 2, 10, 11, 12): pp. 161–186, 187–194, 195–220.

Krishna, S and Badhwar, R.L. (1950). Aromatic plants of India, *J. of Scientific & Industrial Research*, Vol. 9 (No. 3, 4): pp. 221–236, 237–244.

Krishna, S. and Badhwar, R.L. (1952). Aromatic plants of India, *J. of Scientific & Industrial Research*, Vol. 11 (No. 12): pp. 245–246.

Krishna, S. and Badhwar, R.L. (1953). Aromatic plants of India, *J. of Scientific & Industrial Research*, Vol. 12 (No. 2): pp. 267–290.

Kumar, R., Dayal, R. and Sharma, P. (2009). GC-MS analysis of essential oil from aerial parts of *Tagetes minuta*. *Indian Perfumer* 53: 16–19.

Marriott, P., Shellie, R., Cornwell, C. (2001). Gas chromatographic technologies for the analysis of essential oils. *Journal of Chromatography* A, 936: 1–22.

Mendiratta, A., Dayal, R. and Bartley, J.P. (2005). GC-MS analysis of essential oils of needles and twigs of *Cephalotaxus harringtonia* var. *harringtonia*. *J. Essential Oil Research*, 17: 308–309.

Ortiz, F.A. and Robillos, Y.U. (1980). Essential oils in some Philippine plants. *Forpride Digest, IX* (1), 4.

Rashmi and Dayal, R. (2007). Essential constituents of *Achyranthes aspera* leaves. *Indian Perfumer*, 51 (1): 33–34

Singh, V., Dayal, R. and Bartley, J. P. (1999). Volatile constituents of *Vitex negundo* leaves. *Planta Medica*, 65: 580–582.

Singh, V., Dayal, R. and Bartely J.P. (2000). Chemical constituents of volatile oil from *Vitex negundo* flowering twigs. *Indian Perfumer*. 44 (2): 41–47.

Singh, V., Dayal, R. and Bhandari, R.S. (2002). Pesticidial activity of essential oil of *Vitex negundo* leaves against *Sitotroga* cereallela. *Shashpa*, 9(1): 71–75.

Stephen Mial (1940). *A New Dictionary of Chemistry.* Longman Green, London

Varshney, V.K., Dayal, R., Bhandari, R.S., Jyoti, K.N., Prasuna, A.L., Prasad, A.R. and Yadav, J.S. (2005). Behavioral response of the borer beetle *Hoplocerambyx spinicornis* to volatile compounds of the tree *Shorea robusta*. *Chemistry & Biodiversity*, 2785–2791.

Structures of major constituents of essential oils of plant species described in this chapter in above order

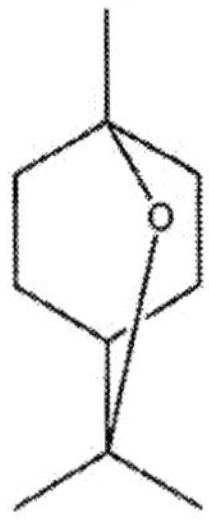

1,8-Cineole

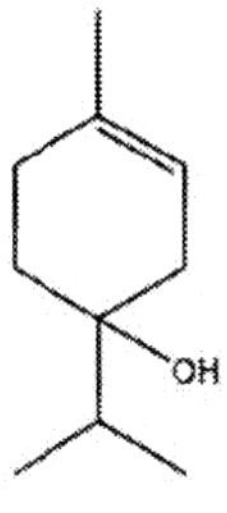

Terpin-1-ene-4-ol

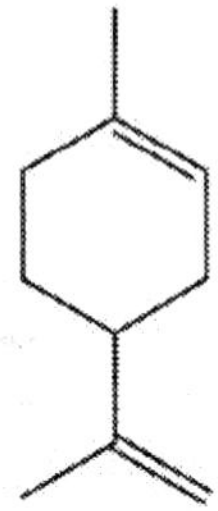

Limonene

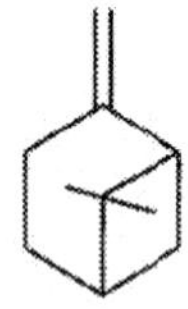

β-Pinene

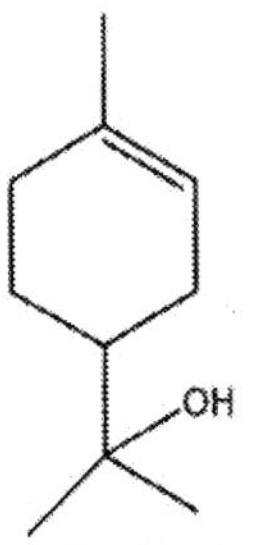

α-Terpineol

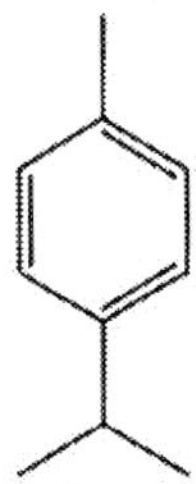

p-Cymene

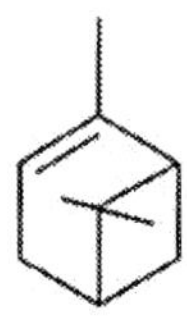

α-Pinene

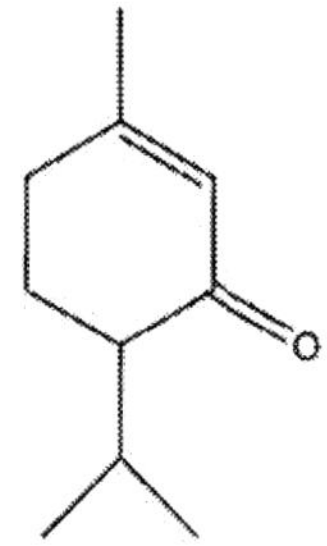

Piperitone

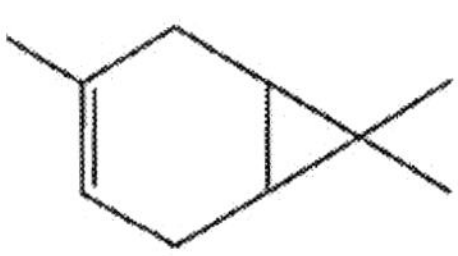

δ-3-Carene

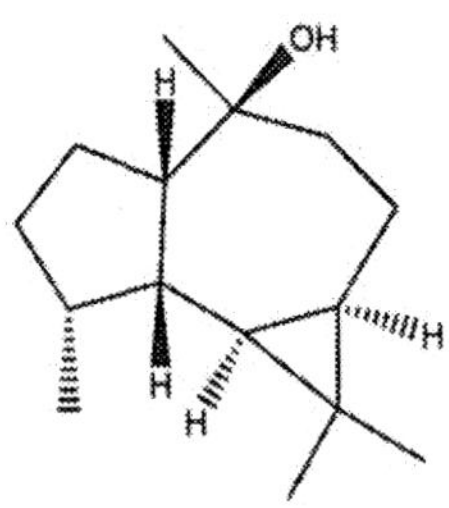

Viridiflorol

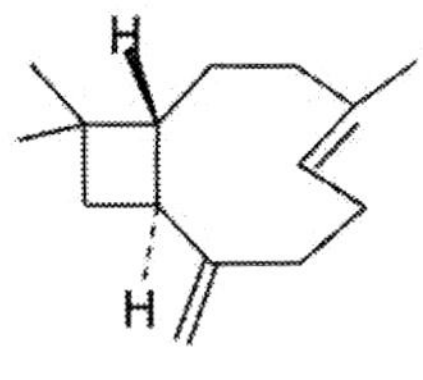

β-Caryophyllene

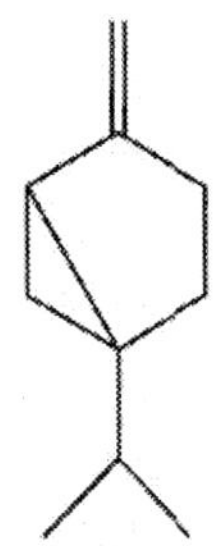

Sabinene

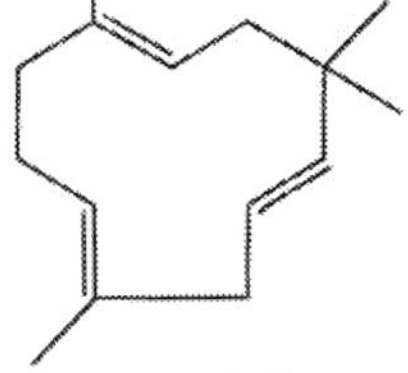

α-Caryophyllene

β-Farnesene

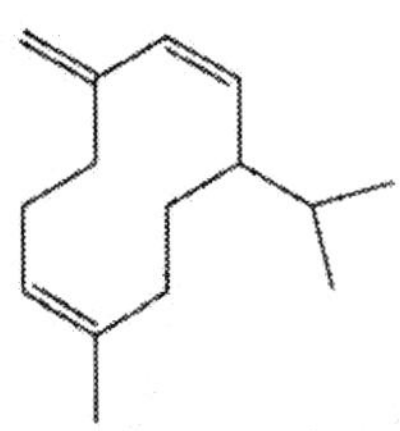

Germacrene-D

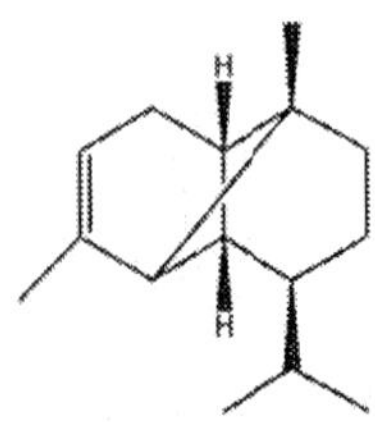

α-Copaene

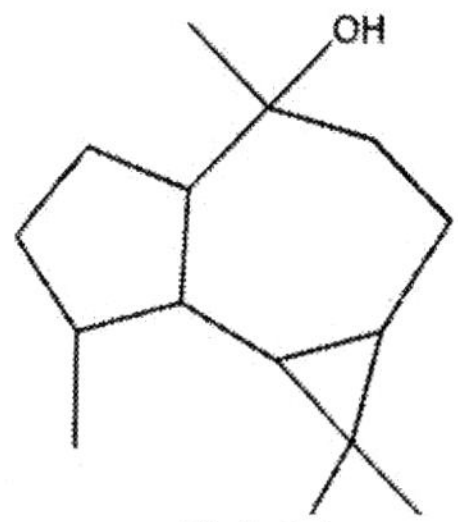

Globulol

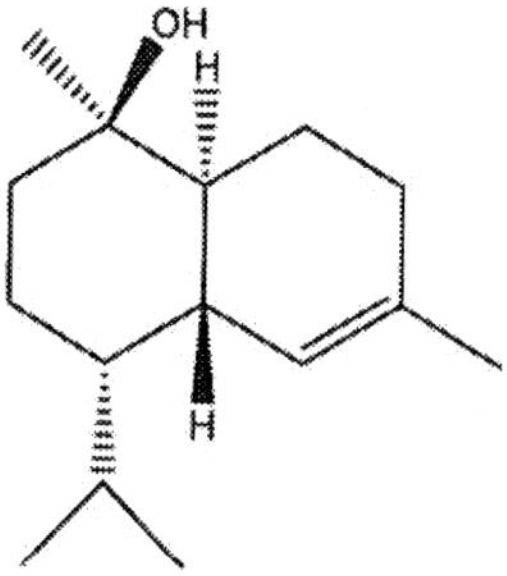

ι-Cadinol

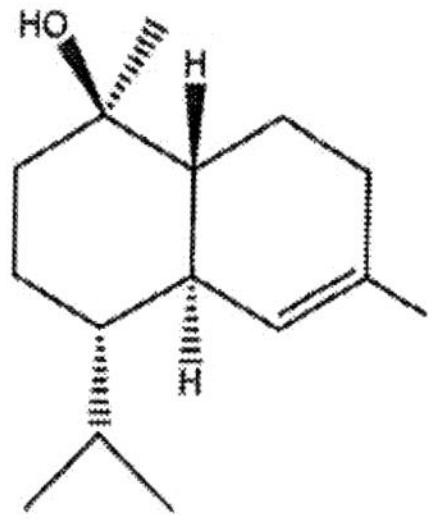

α-Cadinol

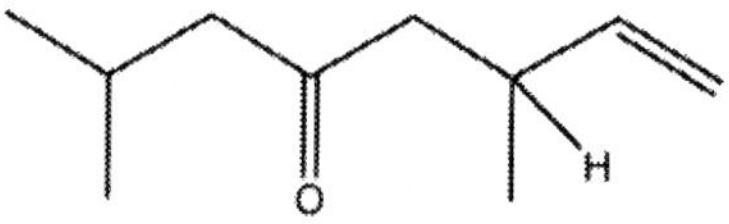

Dihydrotagetone

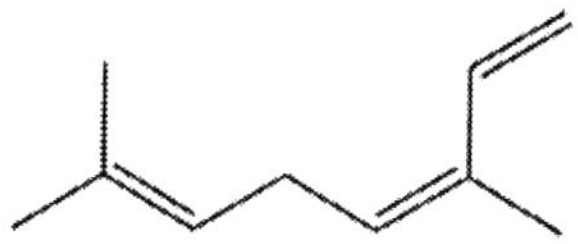

(z)-β-Ocimene

Hydroquinone

7

CHAPTER

Analysis of Fatty Oils from the Seeds of Some Forest Plants

INTRODUCTION

The name lipids includes a class of naturally occurring chemical compounds, which on hydrolysis yield saturated and unsaturated monocarboxylic acids. The lipids are normally soluble in organic solvents, but sparingly soluble in water. The oils, fats, waxes and phospholipids are covered under lipids. Oils and fats are usually known as simple lipids and consist of compounds of glycerol and several saturated and unsaturated organic acids, i.e., they are glycerol esters or glycerides. The difference between oil and fat is that the former is liquid, while the later is solid at room temperature. The oil contains large amount of unsaturated acids than the fat which are richer in saturated acids.

Natural oils and fats occur throughout animal and vegetable kingdom. The oils and fats are present in substantial amounts in the seeds of the plants in association with starch, nitrogenous substance, etc., to assist as nourishment to the seed for germination. All the seeds of different plant species contain oil in varying amount. The seeds containing significant amount of oil can be crushed to isolate the oil are normally known as oilseeds. The oilseeds bearing plants may be classified as annuals and perennials. Groundnut (*Arachis hypogaea*), Sesame (*Sesamum indicum*), Castor (*Ricinus communis*), Linseed (*Linum usitatissimum*), Safflower (*Carthamus tinctorius*) and Mustard/Rapseed (*Brassica* sp.) are the examples of annuals, while coconut (*Cocos nucifera)* and palm are the perennial trees of agriculture origin. Some of the examples of oilseeds of forest origin are from Neem (*Azadirachta indica*), Mahua (*Madhuca indica*), Sal (*Shorea robusta*), Karanj (*Pongamia pinnata*), Kusum (*Schleichera oleosa*), and Jatropha (*Jatropha curcas*) [Lakshmikantan, 1978].

The monocarboxylic acids, present in the glycerides are nearly exclusively straight chain and mainly comprise even number of carbon atoms. The main saturated acids are lauric, myristic, palmitic and stearic acids. The major unsaturated acids are oleic, linoleic and linolenic and normally exist in the *cis-form*. The mixed glycosides are usually present in the oils and fats with the presence of saturated at 1- and 3-positions and unsaturated acids at 2-position. The distribution of acids is not so rigid in the animal fats.

PHYSICOCHEMICAL PROPERTIES OF FATTY OILS

The characterisation of oils and fats is done by determining their physicochemical properties. The physical constants that are usually determined are specific gravity, melting point, refractive index and solidifying point. The type of fatty acids present in the fats and oils can be indicated by some chemical tests namely acid value, saponification value, iodine value, Reichert-Meissl value and acetyl value (Finar, 1973, reprinted 1996).

Acid Value

The acid value indicates the amount of free acid present in the oil or fat. It may be defined as the number of milligrams of potassium hydroxide required to neutralize 1 gram of the oil or fat.

Saponification Value

The number of milligrams of potassium hydroxide required to neutralize acids arising from the complete hydrolysis of 1 gram of the oil or fat is known as saponification value.

Iodine Value

The iodine value indicates the degree of unsaturation of the acids in the oil or fat. The number of grams of iodine that combine with 100 grams of oil or fat is called the iodine value and usually determined by Hubl's or Wij's methods.

Reichert-Meissl Value

The Reichert-Meissl value expresses the amount of steam – volatile fatty acids present in the oil or fat. It is defined as the number of mL 0.1 N-potassium hydroxide required to neutralize the distillate of 5 grams of hydrolysed fat or oil.

Acetyl Value

The acetyl value expresses the number of free hydroxyl groups present in the oil or fat. It is the number of milligrams of potassium hydroxide required to neutralize the acetic acid obtained on hydrolysis of 1 gram of an acetylated oil or fat.

Oils and fats are essential part of human and animal diet and are also of industrial value and find use in soaps, textiles, cosmetics, paper, pharmaceuticals, paints and varnishes, lubricants, biodiesel, etc. The per capita consumption of edible oil is about 11 kg per year in India which is considerably on lower side in comparison to developed countries. Palm oil (mainly imported) and soyabean oil account for nearly half of total edible consumption in India, followed by mustard and groundnut oil. The domestic production of edible oil is from nine primary cultivated oil seeds (groundnut, rape/mustard, soybean and six others) and two major oil bearing material (cotton and Copra). The secondary source consists of solvent extracted oils, rice bran oil, oils from minor and tree-borne oilseeds, etc. Some of the major edible oil being imported in India to meet the gap between the demand and supply are refined palmolein, crude palm oil, crude palmolein, soya bean oil, sunseed oil and other oils. India is the world's fourth largest vegetable oil economy. The industry embraces 15,000 oil mills, 600 solvent extraction units, 250 vanaspati (hydrogenated oil) plants, and over 600 refineries, employing over one million persons. The expected demand of edible oil in 2015 was about 21.3 million tonnes (Mehta, 2006) and now has increased considerably.

ANALYSIS OF FATTY OILS FROM THE SEEDS OF SOME FOREST PLANTS

The fatty oil from the powdered seeds of all the plant species described below was isolated by extraction with petroleum either under reflux conditions using a Soxhlet apparatus. Removal of the solvent (petroleum ether) yielded the fatty oil.

(A) *Cordia dichotoma* Forst. f. syn *C. obliqua* Willd.; *C. myxa* Roxb. Non Linn.

1. **Name of plant species:** *Cordia dichotoma*
2. **Family:** *Borginaceae*
3. **Popular name:** Indian cherry, Lasora (Hindi)
4. **About the plant:** *Cordia dichotoma* is a small or medium sized tree with a short, normally crooked trunk, 3-4 ft in girth. It is extensively distributed in India and Ceylon, especially in warmer area (Anon., 1950). The tree is commonly grown in the land of degraded forests and the one unsuitable for agriculture.
5. **Medicinal and other uses:** The fruits commonly called sebestan, are edible, sweetish, mucilaginous and pickled (Anon., 1950; Anon., 2006). The kernels possess anthelmintic activity. The fruit extract exhibited antibacterial activity against *Bacillus subtilis* and other bacteria. The plant extract forms a component in tropical preparations for skin care/treatment and also possesses the anti-inflammatory and, wound healing activities (Anon., 2006). The floral buds and flowers are used as vegetable after cooking. The rural people of coastal areas in Orissa uses the ripe fruits for eating in raw-form (Anon., 2001, reprinted 2010).
6. **Contribution** of Dayal, Sharma, Sharma and Chandra (Dayal et al., 2006):
   (a) **Fatty oil yield:** 7.60% (Seed Weight Basis)

(b) Physicochemical properties:

Refractive index (28°C)	1.4772
Specefic gravity (30°C)	0.9074
Acid value	50.0
Saponification value	135.3
Iodine value	160.2
% Unsaponifiable matter	1.68

(c) Characterisation of fatty acids in the fatty oil: The fatty acids in the fatty oil were characterized by GLC analysis of the methyl esters. The physico-chemical properties of the oil were determined using standard methods. The oil was saponified with 0.5 N alcoholic potassium hydroxide for 2 hr and mixture of fatty acids was isolated following the normal procedures. Fatty acid methyl esters were prepared by refluxing the mixture of fatty acids with 1 percent methanolic sulphuric acid (H_2SO_4/MeOH) on water bath for 5 hr, cooled and usual work up yielded methyl esters.

The analysis of fatty acid methyl esters was carried out on Chemito Gas Liquid Chromatograph fitted with FID (240°). The temperature of the injector was maintained at 240°C. Capillary Column (25 m, BPX 70, 0.22 mm ID, 0.25 mm) was used using nitrogen as carrier gas (40 ml/min). Split was maintained at 60 ml/min and purge was maintained at 2 ml/min. The oven temperature was programmed from 100–210°C (with increase in temperature 3°C/min) followed by a final hold up time of 20 min. Methyl esters were identified by comparing the retention times with those of standard fatty acid methyl esters and also by their co-injection. The percentages were considered weight percentages (Dayal et al., 2006).

(d) Fatty acid composition:

S. No.	Name of fatty acid	Fatty acid (%) in methyl esters
1.	Pamitoleic acid (16:1)	17.82
2.	Stearic acid (18:1)	26.43
3.	Oleic acid (18.0)	23.12
4.	Linoleic acid (18.2)	26.72
5.	Nonadecanoic acid (C_{19})	2.01
6.	Beheinic acid (C_{22})	1.51
7.	Lignoceric acid (C_{22})	1.52
8.	Unidentified	0.87

7. **Conclusion:** Seven fatty acids were characterised from the fatty oil of the seeds for the first time and linoleic acid (26.27%) was found to be the chief constituent.

(B) *Crataeva nurvala* Buch.-Ham=*C. magna* (Lour.) DC. syn. *C. religiosa* var. *nurvala* (Buch.-Ham) Hook. f. & Thoms

1. **Name of plant:** *Crataeva nurvala*
2. **Family:** *Capparidaceae*
3. **Popular name (Sanskrit):** Marutapaha, Setu vrksha, Varuna
4. **About the plant:** *Crataeva nurvala* is a moderate sized deciduous tree, common throughout India, Burma and Ceylon, either wild or cultivated (Anon., 1950). The multiplication and propagation was tried by culturing axillary buds on MS medium containing sucrose (3%) supplemented with kinetin and BAP and found to be successful (Anon., 2006).
5. **Medicinal and other uses:** Its stem bark is of medicinal value and finds use as a single drug or in compound formulations for the treatment of urinary disorders including urolithiasis, prostatic hypertrophy, neurogenic bladder and chronic urinary infections (Anon, 2001, reprinted 2010). The stem-bark is also widely used as an anti-inflammatory, anti-arithritic and antilithic agent (Anon., 2006).
6. **Contribution** of Dayal, Sharma, Sharma and Chandra (Dayal et al., 2006):

   (a) **Fatty oil yield (Seed weight basis):** 24%

   (b) **Physicochemical properties:**

Refractive index (28°C)	1.4772
Specefic gravity (30°C)	0.9078
Acid value	4.30
Saponification value	181.80
Iodine value	119.20
Unsaponifiable matter (%)	1.05

   (c) **Characterisation of fatty acids in the fatty oil:** The fatty oil was hydrolysed to isolate the mixture of fatty acids which were esterified to obtain the mixture of methyl esters. The mode of characterisation is same as in case of A (Dayal et al., 2006).

(d) Fatty acid composition:

S. No.	Name of fatty acid	Fatty acid (%) in methyl esters
1.	Palmitic acid (16:0)	21.91
2.	Stearic acid (18:0)	14.21
3.	Oleic acid (18:1)	10.30
4.	Linoleic acid (18.2)	36.50
5.	Linolenic acid (18:3)	2.60
6.	Nondecanoic acid (19:0)	0.80
7.	Tricosanoic acid (C_{23})	2.70
8.	Lignoceric acid (C_{24})	1.60
9.	Heptadecanoic acid	0.60
10.	Unidentified	8.78

7. **Conclusion:** A total of nine fatty acids were characterised from the fatty oil of seeds for the first time. Linoleic acid (36.50%) was found to be the chief constituent followed by palmitic acid (21.91%), stearic acid (14.21%) and oleic acid (10.30%).

(C) *Diospyros peregrina* (Gaertn.) Gurke syn. *D. embryopteris* Pers.; *D. malabarica* Desr.

1. **Name of plant species:** *Diospyros peregrina*
2. **Family:** *Ebenaceae*
3. **Popular name:** Gaub Persimmon
4. **About the plant:** *Diospyros peregrina* is an evergreen tree of moderate size and occurs practically throughout India in shady wet places and in the proximity of streams. The trees possess fluted trunk and spreading branches bearing dark green foliage and large velvety fruits. It is an ornamental and shade-bearing tree and thrives on moist soils (Anon., 1952, reprinted 2010).
5. **Medicinal and other uses:** The fruits and stem bark exhibit astringent properties due to the presence of tannins. The unripe fruits are acrid, bitter and oleaginous. The fruits infusion finds use as gargle in aphthae and sore throat. The juice is also used for wounds and ulcers. The seeds oil is used for dysentery and diarrhoea, while bark in dysentery and intermittent fevers. The ether extract of the fruits exhibited anti-bacterial activity against *Escherichia coli*. The wood is hard and heavy and may be used for building construction and boat making (Anon., 1952, reprinted 2010).

6. **Contribution** of Dayal, Sharma, Sharma and Chandra (Dayal et al., 2009):

   (a) **Fatty oil yield (Seed weight basis):** 0.28%

   (b) **Characterisation of fatty acids in the fatty oil:** The fatty acids in the fatty oil were characterized by GLC analysis of the methyl esters. The oil was saponified with 0.5 N alcoholic potassium hydroxide for 2 hr and mixtures of fatty acids were isolated following the normal procedures. Fatty acid methyl esters were prepared by refluxing the mixture of fatty acids with 1 per cent H_2SO_4/MeOH on water bath for 4 hrs, cooled and usual work up yielded methyl esters.

The analysis of fatty acid methyl esters was carried out on Chemito Gas Liquid Chromatograph fitted with FID. The temperature of the injector was maintained at 240°C. Capillary Column (25 m, BPX 70, 0.22 mm ID, 0.25 mm) was used using nitrogen as carrier gas (40 ml/min). Split was maintained at 60 ml/min and purge was maintained at 2 ml/min. The oven temperature was programmed from 100–210°C (with increase in temperature 3°C/min) followed by a final hold up time of 20 min. Methyl esters were identified by comparing the retention times of standard fatty acid methyl esters and also by their co-injection. The percentages were considered weight percentages (Dayal et al., 2009).

   (c) **Fatty acid composition:**

S. No.	Name of fatty acid	Fatty acid (%) in methyl esters
1.	Myristic acid (14:0)	0.90
2.	Palmitic acid (16:0)	20.50
3.	Palmitoleic acid (16.1)	1.10
4.	Stearic acid (18.0)	8.30
5.	Oleic acid (18:1)	25.70
6.	Linoleic acid (18.2)	22.60
7.	Arachidic acid (C_{20})	1.70
8.	Beheinic (C_{22})	4.50
9.	Lignoceric acid (C_{24})	2.60
10.	Unidentified	12.10

7. **Conclusion:** Nine fatty acids were identified for the first time from the fatty oil of the seeds. Oleic acid (25.70%) was found to be the major constituent.

(D) *Hetropanax fragrans* Seem.

1. **Name of plant species:** *Hetropanax fragrans*
2. **Family:** *Araliaceae*
3. **Popular name (Hindi):** Tarla
4. **About the plant:** It is small (30–60 ft) evergreen or deciduous tree distributed throughout sub-Himalayan tract, from Siwalik eastwards to Bihar, Bengal, Assam and in Andaman Islands (Anon., 1959, reprinted 2010).
5. **Medicinal and other uses:** The leaves are used in Assam as a food for eri silkworms as a substitute for castor leaves. Its light brown or grey wood is considered to be useful for articles of turnery (Anon., 1959, reprinted 2010).
6. **Contribution** of Dayal, Sharma, Sharma (Dayal et al., 2007):
   (a) **Fatty oil yield (Kernel weight basis):** 33.1%
   (b) **Physicochemical properties:**

Refractive index (28°C)	1.48
Specefic gravity (27°C)	0.9126
Acid value	5.60
Saponification value	174.40
Iodine value	153.90
% Unsaponifiable matter	2.30

   (c) **Characterisation of fatty acids in the fatty oil:** The fatty oil was hydrolysed to isolate the mixture of fatty acids which were esterified to obtain a mixture of methyl esters. The physicochemical properties of the oil were determined using standard methods (Anon., 1973). The oil was saponified separately with 0.5 N alcoholic potassium hydroxide for 2 hr and mixture of fatty acids were isolated following the normal procedures. Fatty acid methyl esters were prepared by refluxing the mixture of fatty acids with 1 per cent methanolic sulphuric acid (H_2SO_4/MeOH) on water bath for 4 hr, cooled and usual work up yielded methyl esters.

The analysis of fatty acid methyl esters was carried out on Chemito Gas Liquid Chromatograph fitted with FID (240°). The temperature of the injector was maintained at 240°C. Capillary Column (25 m, BPX 70, 0.22 mm ID, 0.25 mm) was used using nitrogen as carrier gas (40 ml/min). Split was maintained at 60 ml/min and purge was maintained at 2 ml/min. The oven temperature was programmed from 100–210°C (with increase in temperature 3°C/min) followed by a final hold up time of 20 min. Methyl esters were identified by comparing the retention times of standard fatty acid methyl esters and also by their co-injection. The percentages were considered weight percentages (Dayal et al., 2007).

(d) Fatty acid composition:

S. No.	Name of fatty acid	Fatty acid (%) in methyl esters
1.	Palmitic acid (16:0)	2.80
2.	Palmitoleic acid (16:1)	0.50
3.	Oleic acid (18:1)	91.20
4.	Linoleic acid (18.2)	2.70
5.	Arachidic acid (20:0)	0.20
6.	Unidentified	2.60

7. **Conclusion:** Five fatty acids were characterised in the fatty oil of the seeds of which oleic acid was found to be the chief constituent (91.20%). The fatty oil can be used as a natural source of oleic acid as it occurs in such a high percentage in it. This is the first report of the isolation and characterisation of fatty oil from *Heteropanax fragrans.*

(E) *Heynea Trijuga* Roxb.

1. **Name of plant species:** *Heynea trijuga*
2. **Family:** *Meliaceae*
3. **Popular name (Kumaon):** Ban-ritha
4. **About the plant:** *Heynea trijuga* is an evergreen tree and attains the height up to 40 ft. It occurs in moist forests throughout, the greater part of India, ascending to an altitude of 6,000 feet. It is also grown as an ornamental tree. The tree possesses dark brown bark with alternate leaves and small fragment flowers (Anon., 1959, reprinted 2010).

5. **Medicinal and other uses:** The leaves, bark and fruits have been reported to be toxic but also have some medicinal value. The greyish wood is moderately hard and heavy and thus used for agricultural implements, beams and scantlings. The fixed-oil isolated from the seeds is used for burning by the Nepalese (Anon., 1959, reprinted 2010).

6. **Contribution** of Dayal, Sharma, Sharma (Dayal et al., 2007):

   (a) **Fatty oil yield (seed weight basis):** 36.16%

   (b) **Physicochemical properties:**

Refractive index (27°C)	1.470
Specefic gravity (27°C)	0.9223
Acid value	10.90
Saponification value	192.20
Iodine value	56.20
% Unsaponifiable matter	0.30

   (c) **Characterisation of fatty acids in the fatty oil:** The mixture of fatty acids obtained after the hydrolysis of fatty oil were esterified to obtain a mixture of methyl esters. The fatty acid composition was determined by Gas Liquid Chromatography (GLC) of the methyl esters by comparing retention time with authentic samples of known methyl esters as in case of D (Dayal et al., 2007).

   (d) **Fatty acid characterised**

S. No.	Name of fatty acid	Fatty acid (%) in methyl esters
1.	Palmitic acid (16:0)	10.40
2.	Palmitoleic acid (16:1)	0.90
3.	Oleic acid (18:1)	62.80
4.	Linoleic acid (18:2)	21.30
5.	Linolenic acid (18:3)	1.50
6.	Nonadecanoic acid (C_{19})	1.20
7.	Arachidic acid (C_{20})	0.50
8.	Unidentified	1.40

7. **Conclusion:** A total of seven fatty acids were characterised in the fatty oil of the seeds. Oleic acid (62.80%) and linoleic acid (21.30%) were found to be the major constituents of the oil. The fatty oil was also transesterified and the

determination of different properties, viz., density, viscocity, CCR value, total sulphur, moisture, copper corrosion, pour point, flash point, cetane index and acid value of the methyl esters indicated that it may be used as biodesel after further studies.

(F) *Lannea coromandelica* (Houtt.) Merrill syn. *L. grandis* Engl.; *Odina wodier* Roxb.

1. **Name of plant species:** *Lannea coromandelica*
2. **Family:** *Anacardiaceae*
3. **Popular name:** Jhingan
4. **About the plant:** *Lannea coromandelica* is a tree of moderate to large size gaining the height up to 24 m and occurs throughout the greater part of India and the Andaman Islands, ascending to an altitude of 1,500 m in the Himalayas. The tree is found in the sub-Himalayan sal (*Shorea robusta*) forests of outer hills and plains, especially in dry places. It is also often planted in avenues and used for hedges and boundaries. The tree grows naturally in adequate quantity and also by transplanting about one year old nursery raised seedlings (Anon., 1962, reprinted 2009).
5. **Medicinal and other uses:** The heartwood is quite durable and used locally for furniture, packing cases, house building and for many other purposes. The mucilaginous gum exuded from the wounds and cracks in the bark is known as *Jhingan* gum. The gum purification can be done by precipitation with alcohol and finds use in paper and cloth sizing, calico-printing and for many other purposes. The gum can be used as a flocculating agent for the purification of sugar cane juice. The bark is acrid and astringent, and is of medicinal value and also yields a strong fibre. It also provides good fodder to cattle, sheep, goat and elephants (Anon., 1962, reprinted 2009).
6. **Contribution** of Dayal, Sharma, Sharma and Chandra (Dayal et al., 2009)
   - **(a) Fatty oil yield (seed weight basis):** 8.70%
   - **(b) Physicochemical properties:**

Refractive index (28°C)	1.453
Acid value	15.2
Saponification value	153.6
Iodine value	17.17
% Unsaponifiable matter	0.94

(c) **Characterisation of fatty acids in the fatty oil:** The physico-chemical properties of the oil were determined using standard methods. The fatty oil was hydrolysed to isolate the mixture of fatty acids which were esterified to obtain the mixture of methyl esters. The mode of characterisation is same as in case of C (Dayal et al., 2009).

(d) **Fatty acid composition:**

S. No.	Name of fatty acid	Fatty acid (%) in methyl esters
1.	Myristic acid (14:0)	0.20
2.	Palmitic acid (16:0)	27.40
3.	Stearic acid (18.0	62.00
4.	Oleic acid (18:1)	6.50
5.	Linoleic acid (18.2)	0.20
6.	Linolenic acid (18:3)	1.80
7.	Unidentified	1.90

7. **Conclusion:** Six fatty acids were characterised in the fatty oil of which stearic acid (62.00%) was found to be the chief constituent.

(G). *Pterospermum acerifolium* Willd

1. **Name of the plant species:** *Pterospermum acerifolium*
2. **Family:** *Sterculiaceae*
3. **Popular name:** Hathipaila
4. **About the plant:** *Pterospermum acerifolium* is a tall evergreen tree up to 24.0 m in height and 2.5 m in girth, having a clean bole up to 12 m, found in sub-Himalayan tract and outer valleys from Yamuna eastwards to West Bengal, and in Assam and Manipur up to an altitude of c. 1,200 m, extending southwards into Ram Nagar hills of Bihar and in Western Ghats of Konkan and North Canara, it is also common in the Andamans. The plant exists in a variety of situations like swamp forests of Dehra Dun, evergreen rainforests of North Kanara and along the riverbanks in the sub-Himalayan tract. It also finds place in the gardens and avenues. It can be regenerated both by seeds and root suckers (Anon, 1969, reprinted 2009).
5. **Medicinal and other uses:** The bitter and acrid flowers are reported to be rich in carbohydrates and render water mucilaginous. These can be eaten raw and are of medicinal value and insect repellent. The leaves find use as platters and for

thatching huts and tobacco packing. The light pinkish red heartwood is not durable in open, but fairly so under cover. The wood is suitable for its use for veeners, panelling, tool handles and for many other purposes. Due to its moderately good bending properties, the wood is used for toys, furniture, walking stick and other ornamental articles (Anon., 1969, reprinted 2009).

6. **Contribution** of Dayal, Sharma, Sharma and Chandra (Dayal et al., 2009):

 (a) **Fatty oil yield (seed weight basis):** 30.0%

 (b) **Physicochemical properties:**

Refractive index (31°C)	1.4773
Specefic gravity (22°C)	0.8934
Acid value	1.12
Saponification value	193.10
Iodine value	92.13
% Unsaponifiable matter	0.24

 (c) **Characterisation of fatty acids in the fatty oil:** The physicochemical properties of the oil were determined using standard properties. The fatty oil was hydrolysed to isolate the mixture of fatty acids which were esterified to obtain the mixture of methyl esters. The mode of characterisation by GLC is same as in case of C (Dayal et al., 2009).

 (d) **Fatty acid compositon (Dayal et al., 2009)**

S. No.	Name of fatty acid	Fatty acid (%) in methyl esters
1.	Myristic acid (14:0)	0.20
2.	Palmitic acid (16:0)	19.00
3.	Oleic acid (18:1)	20.70
4.	Linoleic acid (18.2)	39.50
5.	Linolenic acid (18:3)	1.70
6.	Arachidic acid (C_{20})	1.30
7.	Beheinic acid (C_{22})	6.00
8.	Unidentified	11.50

7. **Conclusion:** Seven compounds were identified in the fatty oil by GLC of the methyl esters. Linoleic acid (39.50%) was found to be the chief constituent. This is the first report of the characterisation of fatty acids in the oil. The fatty acid composition of the oil is very attractive with linoleic acid (39.50%), oleic acid (20.70%) and palmitic acid (19%) as major components and hence the fatty oil maybe a potential source of edible oil after further studies.

Structures of Fatty Acids

Structures of fatty acids characterised from the fatty oils isolated from the seeds of above plant species are given below.

1. Myristic acid ($C_{14}H_{28}O_2$)
   $CH_3(CH_2)_{12}COOH$
   Tetradecanoic acid
2. Palmitic acid ($C_{16}H_{32}O_2$)
   $CH_3(CH_2)_{14}COOH$
   Hexadecanoic acid
3. Palmitoleic acid ($C_{16}H_{30}O_2$)
   $CH_3(CH_2)_5CH{=}CH(CH_2)_7COOH$
   Hexadec-9-enoic acid
4. Margaric acid ($CH_{17}H_{34}O_2$)
   $CH_3(CH_2)_{15}COOH$
   Heptadecanoic acid
5. Stearic acid ($CH_{18}H_{36}O_2$)
   $CH_3(CH_2)_{16}COOH$
   Octadecanoic acid
6. Oleic acid ($C_{18}H_{34}O_2$)
   $CH_3(CH_2)_7CH{=}CH(CH_2)_7COOH$
   Cis–heptadec-8-ene-1-carboxylic aicd
7. Lineloic acid ($C_{18}H_{32}O_2$)
   $CH_3(CH_2)_4CH{=}CHCH_2CH{=}CH(CH_2)_7COOH$
   Heptadec-8, 11-diene-1-carboxylic acid
8. Linolenic acid ($C_{18}H_{30}O_2$)
   $CH_3CH_2CH{=}CHCH_2CH{=}CHCH_2CH{=}CH(CH_2)_7COOH$
   Heptadec–8, 11, 14-triene-1-carboxylic acid
9. Nonadecanoic acid ($C_{19}H_{38}O_2$)
   $CH_3(CH_2)_{17}COOH$
10. Arachidic acid ($C_{20}H_{40}O_2$)
   $CH_3(CH_2)_{18}COOH$
   Eicosanoic acid

11. Behenic acid ($C_{22}H_{44}O_2$)
    $CH_3(CH_2)_{20}COOH$
    Docosanoic acid
12. Tricosanoic acid ($C_{23}H_{46}O_2$)
    $CH_3(CH_2)_{21}COOH$
13. Lignoceric acid ($C_{24}H_{48}O_2$)
    $CH_3(CH_2)_{22}COOH$
    Tetracosanoic acid

REFERENCES

Anon. (1950). *The Wealth of India: A Dictionary of Raw Material Industrial Products. Raw Materials*, Vol. II, pp. 346, 366. National Institute of Science Communication and Information Resources, Council of Scientific and Industrial Research, Publications and Information Directorate, New Delhi, India.

Anon. (1952 reprinted 2010). *The Wealth of India: A Dictionary of Raw Materials and Industrial Products. Raw Materials*, Vol. III:D–E, pp. 76–85. National Institute of Science Communication and Information Resources, Council of Scientific and Industrial Research, Publications and Information Directorate, New Delhi, India.

Anon. (1959 reprinted 2010). *The Wealth of India: A Dictionary of Raw Materials and Industrial Products. Raw Materials*, Vol. V:H–K, pp. 41, 74–75. National Institute of Science Communication and Information Resources, Council of Scientific and Industrial Research, Publications and Information Directorate, New Delhi, India.

Anon. (1962 reprinted 2009). *The Wealth of India: A Dictionary of Raw Materials and Industrial Products. Raw Materials*, Vol. VI:L–M, pp. 27–29. National Institute of Science Communication and Information Resources, Council of Scientific and Industrial Research, Publications and Information Directorate, New Delhi, India.

Anon. (1969 reprinted 2009). *The Wealth of India: A Dictionary of Raw Materials and Industrial Products. Raw Materials*, Vol. VIII:Ph-Re, pp. 308–310. National Institute of Science Communication and Information Resources, Council of Scientific and Industrial Research, Publications and Information Directorate, New Delhi, India.

Anon. (1973). *Official Standard and Recommended Methods, Second Ed*. Society of Analytical Chemists, AOAC, London, 675.

Anon. (2001 reprinted 2010). *The Wealth of India: A Dictionary of Indian Raw Materials and Industrial Products. Raw Materials*, First Supplement Series, Vol. II: Cl-Cy, pp. 201, 232. National Institute of Science Communication and Information Resources, Council of Scientific and Industrial Research, Publications and Information Directorate, New Delhi, India.

Anon. (2006). *The Wealth of India: A Dictionary of Indian Raw Materials and Industrial Products. Raw Materials,* Second Supplement Series, Vol. I:A-F, pp. 235, 241. National Institute of Science Communication and Information Resources, Council of Scientific and Industrial Research, Publications and Information Directorate, New Delhi, India.

Dayal, R., Sharma, M. and Sharma, R. (2007). Fatty acid composition of *Heteropanax fragrans* and *Heynea trijuga* seeds. *Journal of Lipid Science and Technology,* 39(2): 61–62.

Dayal, R., Sharma M., Sharma, R. and Chandra, S. (2006). Analysis of fatty oil of the seeds of *Cordia dichotoma* and *Crataeva nurvala*. *Journal of Lipid Science and Technology,* 38 (4), 183–184.

Dayal R., Sharma M.., Sharma, R. and Chandra, S. (2009) Chemical examination of fatty oils of some forest seeds. *Journal of Lipid Science and Technology*, 41 (1), 8–9.

Finar, I.L (1973 reprinted 1996). *Organic Chemistry, Volume 1,* pp. 321–325, Longman Singapore Publishers (Pte) Ltd, Singapore.

Lakshmikantan, V. (1978). *Tree Borne Oilseeds,* Director, Nonedible Oils & Soap Industry, Khadi & Village Industries Commission, Gramodaya, Irla Road, Vileparle (West), Bombay.

Mehta, B.V (2006). Price outlook and market of palm oil in India. *Sea News Circular,* 9 (3), 25–26.

Index